AF715687

Africa's Lost Wealth

Obert Makaza

Published by Obert Makaza, 2025.

While every precaution has been taken in the preparation of this book, the publisher assumes no responsibility for errors or omissions, or for damages resulting from the use of the information contained herein.

AFRICA'S LOST WEALTH

First edition. November 12, 2025.

Copyright © 2025 Obert Makaza.

ISBN: 979-8232195885

Written by Obert Makaza.

Table of Contents

This book is dedicated to all humanity, who wish to have equality amongst human race.

Africa's Lost Wealth

Obert Makaza:

All rights reserved: Personal opinion

Contents

Chapter One: page: 6

African Soil.

We are missing strong innovative individuals.

The real value of the foreign currencies.

We work for transport and rent.

The education system blinds me from my own wealth.

SA seasons.

Chapter Two: page: 19

Don't sell concepts

Collective circle.

Leakages.

The uncanny ability.

Equatorial Guinea.

Healthy and unhealthy lifestyle.

Personal experience

Challenged by young and fresh bloods

Poor infrastructure negative affect:

Thank you for your service

Loose and leave.

Autocratic government

Rich Africans living abroad

Africans pray to their masters in Europe.

Culture Adaption and Marketing

Brics and dictatorship.

Asia owns parts Africa

Purchase your own identity.

Witchcraft

Introduction

Your curiosity about the title of this book sparks a desire to uncover the secrets of the lost wealth of the world's richest continent. I appreciate that thought-provoking opinions that the title may evoke. It's natural for your mind to seek out robust arguments in defense. In this book, I invite you to temporarily set aside your views on Apartheid and Colonialism. Listen to what I have to say though my views may overflow into the mentioned categories. Continue reading and prepare to be astonished.

Deep down, you're eager to ask, 'What exactly do you mean by African wealth being lost?' You might believe that we possess all we need, yet ownership eludes us. The products and services you use, are not under your name. You are neither the manufacturer nor a shareholder in these commodities.

The dissonance stirred by the book's title is entirely justified. Indeed, the continent remains wealthy, yet the perception of its wealth as 'lost' is tied to investment strategies that have economically marginalized many African communities. The notion that we, as black people, prefer oppression is a harmful stereotype. It's been suggested that if we truly desired change, action would have been taken long ago. **Why invest in a man who dismantles your future, erodes your culture, and undermines your dignity? Why do some Africans silence those who lead with integrity, only to uphold the very colonial systems that once oppressed them?**

Everyone born on African soil is African, regardless of race. The land that first cradled your feet claims you. It is not bloodline or skin tone that defines belonging, but the soil that shaped your first breath. Your elders shape your essence. Language is not inherited through blood; it is gifted by environment. The first word you hear becomes your mother tongue, not because of your skin, but because of your story.

Our systems often act as barriers, their rigidity deterring smart investment choices. With each new system introduced, it seems more low-value goods flood the African black markets. Meanwhile, behind the scenes, the continent's raw materials are quietly extracted to bolster faltering foreign economies.

This is more than a forum, it is a call to action. Here, we forge strategies to reclaim Africa's stolen economic legacy. My mission is not to accuse, but to illuminate. I speak truths in my own voice because I believe that truth, spoken with conviction, has the power to reshape destinies. What you choose to do with what I share in this open discussion is yours alone to decide. But from the very moment you set this book down, the responsibility — and the opportunity — to reshape your future will be in your hands.

Aljazeera

24 May 2017 Nick Deaden article *Africa is not poor, but we are stealing its wealth*

Foreword

My past has imparted important lessons that remain deeply ingrained within me. Among those lessons is the importance of always saying facts. When I speak these truths, I lose people. My words stir anger, provoke discomfort, and sometimes expose the hidden colonial agenda. And almost without fail, I'm told, "You don't know what you're talking about."

This book provides insightful exploration of historical facts. offering a comprehensive understanding of Africa's past and its implications for the future of the continent.

What's even more remarkable is that such organic facts cannot be found anywhere else. The type of governance that allows someone to write and publish something does not always permit the discussion of certain sensitive topics. Some educational systems in Africa today do not have such open discussions due to the intention of breading employees.

Africa's lost Wealth is a remarkable book because it applies to all educational and economic levels, making it accessible and relevant to a wide audience. A lot of individuals have mentioned to me that I am skilled at teaching, indicating that it may not necessarily involve solely lecturing in front of a classroom. I wonder if this might be the most effective way to share what I know. Someone believes that possessing knowledge is crucial, however, I argue that the real importance lies in implementing and utilizing that knowledge.

Many Africans are passing away with their knowledge. Rather than dwelling on the past, we should take action and then provide feedback based on the outcomes. When someone inquired about my actions conflicting with my words, I responded that my book serves as a way for me to practice what I preach.

It is necessary to recover the riches of Africa, but we must acquire them through economic means rather than military intervention. The issue lies within our investment strategies. Altering our strategy towards the type of investment is more important than continuing to adhere to outdated Western systems.

Do not try to accumulate riches as a continent by exploiting religion, as faith cannot be purchased with money. Introducing religion into the pursuit of wealth will only further deteriorate our existing economic condition. Ceasing efforts to solve problems in foreign countries and instead concentrating on our natural resources, while also investing in a highly efficient secondary sector, could significantly boost economic growth in Africa.

Chapter 1

African soil (Alkebulan).

Africa's soil has long been coveted by the world, holding more than 60% of the planet's untapped arable land alongside vast reserves of gold, diamonds, oil, rare earth minerals, and other resources that power global industries. This wealth has made the continent a persistent target for external powers intent on extraction. Yet such interest often comes at a profound cost, disputes over land, minerals, and political control have fueled conflict, deepened internal divisions, and all too often, spilled blood. Tragically, Africans have at times been set against one another, not for the advancement of their own communities, but in service of foreign interests.

The enduring question remains: why must a continent so rich in life and promise be made to bleed for the very abundance that should secure its prosperity?

If you're familiar with African history, you're aware of the continent's vast riches. Before delving into why the world's wealthiest continent is also deemed 'lost.

Copied from Google: *Let's journey back 700 years. It was then discovered that Africa had the most fertile soil on the planet. The 1867 discovery of gold and diamonds in Kimberley sparked European interest, ultimately leading to colonization. Let's take companies like De Beers for instants.*

What was the purpose of De Beers and it's involvement in Apartheid?

NB: (copied from google) *De Beers, a major diamond company, significantly profited from and was intertwined with the system of apartheid in South Africa. While the company's founder, Cecil Rhodes*[1]*, also had colonial ambitions, the apartheid regime, in place from 1948 to 1994, directly benefited De Beers through the provision of cheap labor and a controlled economic environment. Although De Beers also faced some criticism for its labor practices and was involved in initiatives to address inequality, its overall impact during the apartheid era was one of significant financial gain from the system.*

Crucial questions

Why does African soil/ wealth seen as valueless in the hands of the actual owners than in the hands of the foreign investments? Why are we allowing ourselves to be taken advantage of? What exactly our weakness? Why are we fighting amongst us? Until when should we allow those who are contributing to the old colonial system while we're fighting amongst each and losing all the natural resources? Why do we have to get worried when there's a war in foreign countries instead of supporting the ones within Africa? Why are we failing to unite? Is this only because of being greed for money and power or there is something else that we're unable to answer right now, something more valuable than the two categories?

Africa is a continent abundant in natural resources, which should theoretically make it one of the wealthiest regions in the world. However, the value of these primary resources often appears diminished when in the hands of their original owners compared to

---

1. *https://www.google.com/search?cs=0&sca_esv=8d2e7353062bdfd4&sxsrf=AE3TifNgCO3n5S1xSUBebniJCNIVBVwF7Q%3A1754742701673&q=Cecil+Rhodes&sa=X&ved=2ahUKEwjYjJ373f2OAxWvRkEAHRAqNroQxccNegQIBBAB&mstk=AUtExfDTQY3inwgR_uVV3mrig3LkZf1-Xk3kIS7BF54rK3wXcpS8z6YJ44WwPjvha5Dwp__0GRpCv-RjppKXVfK1x45WGr4X2LG2VXvXYm6RJ4Y_gPc1Oib6dYHo4WjvSf5yaBE6Jh9eInlEMFwWJuFyKijRJxMfKGHSCoEdL909Y-EEOvg&csui=3*

foreign entities. This discrepancy can be attributed to several factors, including the perceived value of developed countries' currencies and their economic strategies. Developed nations have historically been able to command higher prices for their currencies, which in turn allows them to acquire primary resources at lower costs, often through the secondary sector. By implementing strategic economic policies and influencing market perceptions, these countries have managed to maintain a competitive advantage.

Consider why African multi-billionaires might be perceived as less affluent compared to foreign investors. Racism remains a significant economic barrier, with the historical legacy of slavery contributing to a persistent undervaluation of darker skin. Being perceived as poor is one thing, but being treated poorly due to such perceptions is a grave and ongoing error. The root of this issue often lies in mindset and self-perception.

Education is pivotal. We need to ask more educational questions and foster a mindset that values our assets appropriately. If one's assets are deemed valueless, there's a tendency to sell at a lower price or abandon them altogether, leading to disinterest in the market and allowing others to advance and that is wrong. Moreover, the current educational system often prepares individuals to be employees rather than economic leaders within their communities. We need leaders, not just employees.

Undoubtedly, there exists a significant distinction between serving the nation and merely being an employee. Many employees unknowingly perpetuate harmful and unsustainable systems that negatively impact society, rather than embracing creativity and contributing to national progress through responsible economic stewardship. By harnessing and safeguarding our most valuable natural resources, individuals have the potential to lead transformative change. However, the average employee often lacks awareness or interest in the long-term future of the nation, limiting their role in shaping a more equitable and sustainable society.

We must continue posing these critical questions. The more we interrogate the issues surrounding Africa's lost wealth, the closer we come to uncovering meaningful solutions. In many of our communities, we often default to hypothetical thinking—asking "if" rather than actively seeking the best path forward. To address this challenge effectively, we must first identify who holds the wealth we seek. Understanding the individuals or entities in control of these resources is a vital step toward reclaiming and redistributing them for the continent's benefit.

To address these challenges, it's crucial to reform the educational system to emphasize leadership and economic empowerment. By doing so, we can cultivate a generation that not only values their assets and heritage but also possesses the skills and knowledge to lead their societies towards economic prosperity. We should stop implementing educational systems that breads employees. This type of educational system is more harmful than it initially appears. Beneath its surface, it often perpetuates outdated frameworks, suppresses critical thinking, and limits the creative potential of learners. Without meaningful reform, such systems risk undermining societal progress and failing to equip individuals with the tools needed to address complex national and global challenges.

Africa's greatness will not be built by education alone; the conduct of white people must change as well. Liberate the Black person, and you will discover abundance beyond measure. In freeing Black people, you free the Earth itself, allowing it to heal. History shows that capitalism has long been sustained by white privilege, and that religions such as Christianity have often flourished through the oppression of Black people. So yes, free Black people, and you free the world, opening the door to a future where all can live in happiness and plenty. We all know that Black people carry a unique DNA that white people depend on for survival. So why not protect this valuable DNA to ensure it remains strong and healthy?

We are missing strong innovative individuals

We require courageous individuals who are ready to embrace calculated risks to significantly advance the African economy. There is no necessity for imported tertiary goods and services, nor is there a need for military intervention to reclaim wealth, we do not seek conflicts against advanced first world countries. Instead, we should employ the most effective business strategies to eradicate economic disparities within our continent. True leadership is characterized by the ability to devise strategic solutions that secure and enhance valuable assets.

The wealth of Africa, often characterized by its abundant natural resources, remains largely under the control of foreign investments rather than in the hands of its citizens. Despite gaining independence from colonial powers nearly two to three decades ago, some African countries have not shown substantial economic improvement. This stagnation is not due to a lack of resources but rather to systemic issues that create barriers to progress. Poor leadership has been a critical factor in these circumstances. The remnants of colonial systems still influence approximately 60% of the African continent. Zimbabwe, for example, gained independence earlier than many other African nations, including South Africa. However, today, Zimbabwe is among the poorest countries, which can be attributed to economic mismanagement.

We are not only marginalizing the real leaders in Africa, but we are also silencing them. Time and again, great leaders who propose viable solutions to economic challenges meet untimely ends. Instead of uniting under a single economic vision, we find ourselves divided by multiple political agendas. This division is one of the reasons why our economies are overshadowed by foreign nations, which often present a united front. This unity enables them to access Africa's raw materials more easily. These primary raw materials are purchased at low costs, processed abroad, and then returned to us as finished products at a much higher price. Moreover, the import taxes imposed on these goods and services, which are derived from the very raw materials we exported, only add to the economic burden.

Why did the CIA allegedly send Carl Colby to give Bob Marley a pair of shoes containing a needle after which he developed cancer only weeks later? What was the real motive behind their actions? Even before that, they had reportedly sent two armed agents to Marley's home in an attempt to kill him by gunfire. Why? And why did they also attempt to kill Maxwell Chikumbudzo through food poisoning? Does darker skin threaten the world? And if so, in what way could we possibly be a threat to the rest of the world?

The real value of the foreign currencies.

Foreign currencies do not have an intrinsic base value like physical commodities such as gold, diamonds, or oil. However, some currencies are indirectly linked to these natural resources through various mechanisms. (*You'll have to forgive me, I don't fully understand how markets work. Just like I was once told, "You don't know what you're talking about." And maybe she was right. I need to go back to school and learn how to write properly)* Foreign currencies do indirectly depend on commodities, their stability and value are influenced by factors such as gold reserves, natural resources, and economic policies. Proximate 30% of the world entire minerals reserves are in Africa.

In the television series *Mr. Robot,* the character Elliot challenges the glorification of figures like Steve Jobs, arguing that such heroism is undeserved when built on exploitative foundations. He specifically references the use of child labor in cobalt mining in the Democratic Republic of Congo, a critical component in the production of modern electronics. It is estimated that approximately 30% of mine workers in Congo are children, many of whom receive no compensation unless they happen to work under someone who shows genuine concern for their well-being. Some children labor solely to support their families, enduring harsh conditions with little protection or recognition. These young workers often walk barefoot, go to bed hungry, and live in communities scarred by uncovered mining pits. Meanwhile, the families of corporate elites who profit from this supply chain enjoy lives of luxury, rarely acknowledging the human cost behind their wealth. It is a sobering truth: those who contribute most to the creation of global goods and services are often the least likely to benefit from them, their labor undervalued, and their humanity overlooked.

Given that these currencies lack significant baseline value, the decision was made to derive value from cryptocurrency instead. The strategy initially posed significant challenges in assessing and attributing value to their currencies. When cryptocurrencies emerged, both local and foreign leaders expressed skepticism and hesitancy in trusting this new form of currency. Currently, there is a growing understanding of how cryptocurrencies may impact the economy either negatively or positively.

Do you understand where crypto currency came from? Now, let's refer you back to the paragraphs above. As African countries impose more sanctions on the outside world, new currencies are emerging as a result. In Africa, when inflation rises, youth are often the most vulnerable group, facing severe food insecurity even more acutely than adults. Our elders possess a deeper understanding of life than we often do, despite our upbringing in an era defined by advanced technology. Growing up amid what appears to be the triumph of the technological era does not mean we possess complete understanding or knowledge.

We often forget that it was they who played a pivotal role in shaping the very foundations of what now appears to be a luxurious modern lifestyle. It was through their hardship that some of them came to harness wealth and more enduringly, a deep and intuitive grasp of economic dynamics.

During periods of economic strain, governments often target the middle class, urging them to pay higher taxes, encouraging businesses to contribute more to charities, and emphasizing investment in community development. Regrettably, many of these economic strategies find their origins in Western practices or were imposed during colonial eras. It's ironic that some government officials advocate for businesses to contribute to community development while failing to do so themselves.

We work for transport and rent.

Does such economic burden get down to the last bit of your buck? It's a common observation that many individuals living in urban environments may not fully realize the lives they aspire to lead. The urban lifestyle often comes with its own set of challenges, including high costs of living, increased competition, and societal pressures. It's not uncommon to witness life unfold in an unexpected ways, where our aspirations and reality may diverge. It can be incredibly frustrating when you're on the verge of exploring business opportunities, only to encounter unexpected obstacles such as new laws and potential crime. In many African countries, the dynamics of urban life present unique challenges, such as working for transport and rent and the rest mere cents goes back home to the families in rural.

The struggle against oppression continues, albeit in evolving forms. Understanding the diverse struggles faced by people across Africa requires immersing oneself in the continent's rich tapestry of cultures, histories, and lived experiences. I underestimated my experience of leaving Zimbabwe and encountering different types of life elsewhere. As my mother once said and she was right. I might have died young or become a polygamous father had I grown up in Zimbabwe.

I'm blinded to my own wealth.

The landscape of drugs and their impact on individuals and societies is multifaceted. While some substances are widely recognized as harmful, others may surprise us. That drug is not one that can be measured in milligrams or detected in urine tests. Instead, it permeates our societies, infiltrating institutions, minds, and hearts. Traditions within households often transcend borders, cultures, and generations. These shared practices weave the fabric of our lives, connecting us to our roots and shaping our sense of identity. The traditional life path I have like birth, education, career, retirement, and then paycheck, has been deeply ingrained in societal narratives. A significant percentage of the world's population adheres to this familiar script especially Africans. The intricate dance between medication and its side effects mirrors the delicate balance within our societal systems. It is indeed disheartening that Africans have historically borne the brunt of slavery. I refrain from repeatedly using that word; instead, I am simply stating a fact as a person of Egyptian Nigro descent.

The challenges faced by many in Africa can be deeply disheartening. While the continent boasts rich cultural heritage, breathtaking landscapes, and resilient communities, it also grapples with significant hardships. The historical legacy of colonialism and its impact on Africa's socio-economic landscape is profound. It is indeed a poignant reality that some individuals, despite their educational achievements, find themselves facing immense challenges and, tragically, may pass away without fully realizing the potential that their certificates, degrees, or diplomas represent.

What happened in the above paragraphs? Picture two smart gorillas. One Gorilla excels as a hard worker in physical tasks, while the other stands out for having the best ideas. Now these two Gorillas strives to achieve something that would significantly enhance their quality of life. The intellectually astute gorilla brings forth a brilliant idea during the discussion, but recognizing their own physical limitations, they graciously pass the torch to the physically fit counterpart to execute the plan on their behalf. **Solution:** We should stop saying that white people are smarter when we facing systemic challenges. It is crucial to move beyond simplistic associations tied to racial groups. Every race is smart and everyone else smart in their ways. Despite one race keeps getting ahead, you need to look for opportunities that are available today. I understand the limitations these systems create, but every challenge gives birth to new opportunities. So, how willing are you to learn from the challenges these systems present?

SA seasons:

South Africa is known for its diverse climate and unique seasonal variations. In addition to the typical four seasons (spring, summer, autumn, and winter), South Africa experiences some additional climatic phenomena that make living there quite interesting. I'm sorry if this sounds amusing, but unfortunately, it's the truth.

These are the extra seasons of South Africa that many of the outside world do not know and or might have misunderstood:

1) In South Africa, the striking season often reveals a troubling pattern: it is not uncommon for businesses and organizations to fail in

compensating workers fairly. In response, workers frequently resort to protests to voice their grievances—yet these demonstrations are not always conducted constructively. Rather than remaining confined to a single organization, such protests can spill over into broader societal unrest. Factors like unresolved grievances, excessive taxation, and rising costs of goods and services further fuel collective frustration.

What's both concerning and paradoxical is how some protests result in the destruction of essential services while simultaneously demanding their improvement. Imagine a community so desperate for access to education that it burns public transport in protest. Then, due to delays in resolving the transport issue, perhaps because of insurance or bureaucratic hurdles, the same community turns to destroying schools in demand for restored transportation. This cycle of reactive destruction underscores a deeper systemic failure and highlights the urgent need for more thoughtful, sustainable forms of civic engagement.

2) **Xenophobia seasons:**

They plant the seeds of fear in spring,

Harvest hatred come the fall.

Yet every season, truth still sings

The earth was never theirs at all.

African Xenophobia victims of violence or discrimination experience targeted attacks based on their ethnicity, nationality, or race; they often refer it as Afrophobia. Incidents of violence or discrimination are not isolated acts carried out randomly by individuals within communities. Such types of social issues are a very complex issues with deep-rooted causes. This behavior is often exhibited by individuals who choose not to work but instead seek to benefit by causing harm to innocent people. It is often associated with individuals who exploit government systems to their advantage, causing harm to both none citizens and citizens.

Allegedly, among country leaders, there is an orchestrated annual event aimed at eliminating a specific number of both documented and undocumented foreigners. Interestingly, there are instances where undocumented foreigners are also involved in executing harmful actions against their fellow migrants. Some individuals have resorted to participating in robbery activities, while others employ kidnapping as a survival strategy. In certain townships, young people sometimes feel unsafe walking alone due to concerns that either wealthy individual regardless of their race may be involved in illegal organ trafficking.

What is Dudula Operation and who started Dudula operation in South Africa? Why does it seem to target one set of foreign nation? (NB: **this is copied from Wikipedia**) Operation Dudula was founded by Nhlanhla "Lux" Dhlamini (also Mohlauhi). Dlamini rose to prominence for his role in defending Maphonya Mall from looters in the July 2021 unrest. The group emerged from discourse that blamed migrants for the fallout and economic hardship of COVID-19 deaths and lockdowns. Operation Dudula led their first march on 16 June 2021 through Soweto targeting people they that they believed were foreign drug traffickers and businesses that they thought employed immigrants. The march increased their popularity and in the following months, several other anti-immigrant groups also going by Dudula or some variation of the name, such as the separate Alexandra Dudula Movement, were established. In April 2022, Operation Dudula expanded to Durban, KwaZulu Natal, and in May 2022, to Western Cape Province.

Operation Dudula states that its campaign is aimed at addressing crime, a lack of jobs and poor health services caused by an "influx of illegal immigrants". They have campaigned for small businesses to only employ South Africans, and for migrant shop keepers to close .down and leave South Africa. The campaign has been accused of a number of instances of violence against African immigrants in South African townships, including forcibly closing shops and raiding properties.

Mmusi Maimane and Julius Malema said that; Document those who are undocumented. Maine and Malema's point was, they see that the people who mostly treated bad here, are both legal and illegal immigrants leaving in South African.

Obert's opinion (I believe the two gentlemen raised strong and valid points.) Maimane went further by validating his point saying that all he doesn't want is the little foreign nations within South African cities isolated. We must engage and learn from each other and grow the economy together. He said that people must interact not remain divided. The more we continue fighting amongst ourselves the more we remain divided, and the more colonial systems are happy, as they will use that opportunity to take over.

Is it true that the migrants are solely the cause of economical hardships in South Africa? Is there anything the government is doing to stop the foreign influx? (NB: **this is copied from Wikipedia**) Operation Dudula has been termed Afrophobic. Supporters of migrant rights argue that Operation Dudula is unjustly blaming migrants for economic hardship caused by South Africa's deep inequality.

We currently do not hold any evidence that Zandile and Gayton Mckenzie to be originally Zimbabwe as the social media claimed. What leads the public to say that they both seem to be originated from Zimbabwe?

NB: (copied from google) Gayton part: Read: Meet South Africa's GNU's Cabinet: "McKenzie intentionally, for admitted political reasons, communicated his desire to harm foreign nationals receiving treatment in South African hospitals in a manner that could cause their death, in public broadcasts on national television," says the complaint. (It adds that this violates the Prevention and Combatting of Hate Crimes and Hate Speech Act).

McKenzie's 2022 threat to personally switch off foreign nationals' oxygen machines in hospitals was widely quoted in the South African press. This came shortly after a video of then Limpopo Health MEC Phophi Ramathuba , who is now premier of the province (since June), verbally attacking a Zimbabwean patient for coming to SA rather than staying in Zimbabwe for treatment, claiming that foreign nationals were overburdening the healthcare system.

McKenzie later told eNCA: "If there is a South African, Zimbabwean and Mozambican patient on oxygen and I see a South African patient born and bred in SA; I will turn the oxygen off so that the South African can live." This prompted the Progressive Health Forum (PHF) to call for criminal sanctions against those who incite murder. PHF convener Dr Aslam Dasoo told News24 that both Ramathuba and McKenzie's comments were the language of ethnic cleansing and genocide.

If Mr. Gayton McKenzie were originally from Zimbabwe, could he truly despise his own people? Public speculation about his heritage raises deeper questions: if his parents did migrate from Zimbabwe to South Africa, what forces might have driven that decision? Could it have been political exile, economic hardship, or even displacement under Rhodesian rule?

And what of Ms. Zandile Dabula? If her personal conflicts with the father of her children have shaped her views, is that pain now being weaponized against innocent foreigners, particularly Zimbabweans? In a troubling twist, many supporters of Operation Dudula appear to be foreign nationals themselves. What does that say about the movement's true nature?

Is this not a modernized tribal civil war, masked by slogans, fueled by unresolved trauma, and perhaps orchestrated to serve the interests of upcoming elections, money, and power? Beneath the surface of patriotism, might we be witnessing a deeper betrayal: Africans turning on Africans, while the architects of division remain unseen?

The actions of this newly formed informal political party are deeply harmful to the progress of our society. While we remain distracted by their destructive behavior, colonial interests are watching and quietly seizing opportunities in the chaos. We must all zoom out from the chaos unfolding across the country if we are to pursue a shared prosperity as a united African collectively.

3) **Loadshedding season:**

In 2019 during my visit to Zimbabwe, I attempted to obtain my documents, but I was informed that I couldn't be considered a Zimbabwean since I failed to speak Shona. In a public setting, I was loudly told by someone to return either to South Africa or my country of origin. That's a story for another day. Please read LIFE TIME RUNS (biography). NB: I am not claiming to be entitled to remain in South Africa, however coming to South Africa was such a great, saving my life by running away from an abusive family was not a choice. (South Africans I am so grateful, and I will forever be grateful.)

To use my cellphone, I often had to wait until 23:00 when the power would briefly return only for it to go off again by 02:00. I understand how deeply disruptive load shedding can be, especially when it affects daily life and essential services. While South Africa has faced significant power challenges, Zimbabwe's situation has been even more severe. Power cuts there often last over 12 hours a day, driven by outdated infrastructure, low generation capacity, and widespread mismanagement.

As for the claim about Congolese individuals obtaining documents in South Africa due to coal sales, there's no verified evidence supporting a direct link between Congolese coal exports and immigration documentation. In fact, Congo's coal exports to South Africa are minimal, with only $26 worth recorded in 2017. The broader trade relationship includes minerals and metals but not coal at a scale that would influence immigration policy.

South Africa's resource imports play a vital role in sustaining its economic health, and trade with other African nations is a cornerstone of regional integration and global interconnectedness. This dynamic has sparked deeper conversations about South Africa's complex relationship with the rest of the continent, one shaped by historical ties, economic disparities, and evolving political alliances.

African American or Black American.

"*You are just like us but you're not one of us*". I first encountered this perspective while staying with someone in Khayelitsha. During a recent visit from a group of Americans, a friend of mine repeatedly underscored the significance of the conversations we shared. Some of the visitors expressed a desire to distance themselves from broader racial affiliations, revealing complex layers of identity and self-perception.

Contrary to some explanations, our experience spending two weeks with them during the visits to volunteer and revitalize basketball courts in Khayelitsha in 2018 couldn't be further from the truth.

As much that I like money, I have been wondering why African Americans do not invest in Africa? Experts have explained that the reluctance of wealthy individuals to invest in their countries() of origin often has less to do with personal wealth and more to do with systemic issues such as political instability, corruption, and lack of infrastructure. While beliefs in black magic and secret societies like the Illuminati exist in African cultural narratives, there is no verified evidence linking these to the investment decisions of wealthy individuals (**what is Illuminati?** The Illuminati (plural of Latin[2] *illuminatus*, 'enlightened') is a name given to several groups, both real and fictitious. Historically, the name usually refers to the Bavarian Illuminati, an Enlightenment[3]-era secret society[4] founded on 1 May 1776 in the Electorate of Bavaria[5].

What was Adam Weishaupt's intentions of starting Illuminati?

The society's stated goals were to oppose superstition[6], obscurantism[7], religious influence over public life, and abuses of state power[8]. "The order of the day", they wrote in their general statutes, "is to put an end to the machinations of the purveyors of injustice, to control them without dominating them." The Illuminati, along with Freemasonry[9]

2. https://en.wikipedia.org/wiki/Latin
3. https://en.wikipedia.org/wiki/Age_of_Enlightenment
4. https://en.wikipedia.org/wiki/Secret_society
5. https://en.wikipedia.org/wiki/Electorate_of_Bavaria
6. https://en.wikipedia.org/wiki/Superstition
7. https://en.wikipedia.org/wiki/Obscurantism
8. https://en.wikipedia.org/wiki/Power_(social_and_political)

and other secret societies, were outlawed through edict[10] by Charles Theodore, Elector of Bavaria[11], with the encouragement of the Catholic Church[12], in 1784, 1785, 1787 and 1790. During subsequent years, the group was generally vilified by conservative and religious critics, who claimed that the Illuminati continued underground and were responsible for the French Revolution[13]. It attracted literary men such as Johann Wolfgang von Goethe, [14]Johann Gottfried Herder[15] and the reigning Duke of Gotha[16] and of Weimar[17].

How is the illuminati affecting Africa today? Answers: Clement Majawa from the Catholic University of Eastern Africa; he said: *The Order of the Illuminati is often at the center of debates about the impact of this secret society in Church and society today. Is the Illuminati a myth or does it truly secretly rule the world? Why is this secret society of Illuminati famous and yet no one comes out in the open and explain its existence or show allegiance to it. Maltese Bishop Late Emmanuel Barbara (1949-2018) of Malindi diocese addressed a conference at Don Bosco Utumi Theological college in Karen, Kenya. He revealed that Illuminati is a challenging pastoral issue which threatens the growth of faith and prevents deeper evangelization because research has shown that the Bishops in Africa are aware of such secret societies or groups, and how it entice Christians to join such occults for prosperity, fame, monetary beneficiaries and become influential in society. Yet, agents of evangelization are uncomfortable to discuss it; local catechisms or*

9. https://en.wikipedia.org/wiki/Freemasonry

10. https://en.wikipedia.org/wiki/Edict

11. https://en.wikipedia.org/wiki/Charles_Theodore,_Elector_of_Bavaria

12. https://en.wikipedia.org/wiki/Catholic_Church

13. https://en.wikipedia.org/wiki/French_Revolution

14. https://en.wikipedia.org/wiki/Johann_Wolfgang_von_Goethe

15. https://en.wikipedia.org/wiki/Johann_Gottfried_Herder

16. https://en.wikipedia.org/wiki/Duke_of_Gotha

17. https://en.wikipedia.org/wiki/Duke_of_Weimar

teachings overlook the topic of illuminati. The Illuminati is dangerous at all fronts of the Church and society because initially, it was meant to be a secretive organization (1776-1785) built on gnostic ideas and began by a German philosopher and theologian Adam Weishaupt. It was meant for enlightened intellectuals whose goal was the overthrow of an oppressive monarchy and the control of civil authorities by the Roman Catholic Church. Often it is credited for influencing the French Revolution in 1787. The Illuminati taught a doctrine of social revolution and political liberation that hinged on the equality of man and woman, the embrace of relativism and rationalism, and the denial of powerful authority of the Church considered as the legitimate institutions for the regulation of social and moral values in the world)

Obert's opinion: *Based on Clement's explanation, it's important to recognize that the responsibility to "fix" or "grow" Africa doesn't rest solely on our shoulders. When some of our leaders are affiliated with secretive organizations, it raises questions about who truly holds influence over the continent. Historically, groups like the Dutch Broederbond in South Africa wielded immense power behind the scenes, shaping national policy and leadership for decades.*

The concern is that when presidents or key figures belong to such secret societies such as Illuminati, it can undermine transparency and hinder genuine progress. This level of control often lies beyond the reach of ordinary citizens. We tend to fight among ourselves, yet the deeper truth is that we are not each other's true enemies. The real disruption stems from those who sow division and confusion for their own gain.

What happened to Akon's idea of building futurist city in Senegal?

NB: (Google answers maybe incorrect) Akon *first publicly announced his plan to build a futuristic city in Senegal in June 2018. Specifically, this announcement was made during the Cannes Lions International Festival of Creativity. The city, named Akon City, was initially envisioned as a*

$6 billion project featuring Wakanda-style skyscrapers and running on the singer's cryptocurrency, Akoin.

Well sounds good isn't it, how far has he gone then?

Please allow me to say that Akon's ambitious vision for Akon City now appears to be fading into history. Several investors have come forward claiming they were misled and are demanding the return of their funds. Despite the initial excitement, there's been little visible progress on the project. Allegations of fraud have surfaced, and Akon himself has remained largely out of the public eye regarding the matter.

No one truly knows the reason behind the setbacks.

But if it had been a foreign investor, confident in the promise of massive returns, I'm certain we'd already be en route to Senegal. NB(**copied from google**) The project has been abandoned by both Akon and the Senegalese government. Most of the 800-hectare site, located 100km south of Dakar, remains undeveloped. While some initial construction, including a welcome center, was started, the project has faced numerous setbacks and is now being reclaimed by the Senegalese government for a tourism hub.

Akon City may have once stood as one of Africa's boldest economic visions, but it's far from the only one that's been sidelined. Trust me, there are greater minds we never got to meet, they probably silenced before they could speak. Visionary leaders are being told to step aside, not because they lack merit, but because their ideas threaten the personal gain of well-known figures. And instead of uplifting African societies, these projects serve individual interests.

It's especially tragic when governments are complicit because then, their power becomes not a tool for progress, but a symbol of corruption and decay.

The question remains: *How long will we allow corruption to dictate the market? Is it true that, without corruption, some households would have nothing to eat? Basically without Black Market you wouldn't been able to read this book.*

Either way, this system is stifling our true potential.

Chapter Two

Don't sell us concepts.

In Africa, we often engage in discussions but seldom follow through with action, which underscores the significance of foreign investment. Many foreign investors are driven by clear objectives and the ability to execute plans effectively, capitalizing on the strength of their currencies to achieve their goals. For example; many do not know why Germany is the most feared when it comes to weapons. (*Albert Einstein advised the U.S. government never to allow Germany to rearm, warning that their history of militarism posed a serious threat to global peace.*)

In sectors such as transportation, agriculture, technology, medicine, and education, we actively encourage foreign investments to take a leading role. The reluctance to embrace risk is a key factor in our dependency on foreign investment. Additionally, the efficacy of government tax laws is undermined by the pervasive challenge of corruption, which, despite being a matter of choice, affects all levels of society. This is the moment we see it clearly: billionaires, not governments, are running the show. Is this a strategic game among elites, where one says 'make your move,' and the other responds, 'I'll follow with mine,' as they compete to see who truly controls society? For me, this is a cruel yet fascinating kind of civilization we've entered. Some of the most advanced thinkers have long dreamed of reaching this stage and now, at last, it is happening. So, does that mean they feel happy and scared at the same time?

We all seek to create lasting memories, but their significance can vary from person to person. Consider the tourists with cameras in hand; they are not merely capturing images. They represent a broader interest in seizing business opportunities that could otherwise belong to local entrepreneurs. How do they manage it?

Listen closely: I am trying to teach some you that, not all visitors are here for leisure. Among them are individuals dispatched by their governments, corporate leaders, and elite investors, tasked with exploring African lands for potential investment opportunities. While we are engaged in protests, burning tyres on the streets, these foxes are diligently surveying areas with lower crime rates to establish new businesses. Of course, we all understand that crime can never be completely eradicated. However, with the right business strategies, it is possible to significantly reduce certain types of crime, particularly those beyond white-collar offenses. Sometimes, you shouldn't rush to explain your culture to these travelers. We've seen it in movies, they first study the culture to find its weaknesses, then move in to take over someone else's land and resources. So why don't we learn from that?

That's why certain businesses won't let you take photos inside. They recognize that you may be able to identify business opportunities that have escaped their notice.

All these types of strategies could have been implemented by us, rather than relying on foreign entities. Today, these very foreign entities are the ones who own, operate, and implement the systems we are expected to live under. Seen from another perspective, it's like pinching a child until they cry, then offering them sweets and water to soothe the pain. Over time, the child becomes conditioned to expect a reward each time they're hurt, gradually adapting to a system where comfort is used to mask the cause of their suffering. So does the child perish in this suffering, only for their descendants to inherit the same broken and misleading systems?

In many nations still governed by royal families, the focus tends to be on serving the interests of the monarchy. The term 'king' evokes mixed reactions: for some, it symbolizes leadership; for others, it connotes dictatorship or selfishness. Personally, I find the notion of selfishness most resonant.

I recall viewing a documentary where an Eastern monarch distributed funds to his subjects in a way that seemed patronizing, reminiscent of tossing scraps to a famished dog. The sight was poignant: the recipient knelt as though receiving a treasure, yet it was merely a small sum, insufficient even to purchase a loaf of bread. Even today, I still witness the same patterns in nearby communities, wealthy individuals tossing spare change to trolley workers, car guards, or shop cashiers with a sense of arrogance, as if generosity were a performance rather than a gesture of dignity.

To gain a deeper understanding of the cultural and royal dynamics, one might consider visiting various African royal palaces to observe the customs and interactions of the people. People often spend time researching the wealth of others, inadvertently neglecting their own goals. Why should I throw away someone else's food and clothes, and tell them they're not welcome in my area or country, just because of their economic status? Australia's borders are known for treating international travelers this way.

Social and my time.

Time flows like a river,

its waters never the same.

Spend each drop with purpose,

it's the richest wealth you'll claim.

I, too, still sometimes engrossed in videos of wealthy individuals, which do leads to a realization of the precious time I squander. Although I often download countless videos to my phone, they sometimes lead to procrastination instead of progress toward my aspirations. While writing this book, I was actually taking a break from the mental strain of trying to deeply understand trading, immersed in YouTube videos from Ahmed XM and Habby Forex, where success is flaunted through flashy cars and boastful mentions. I still have a female friend who always says "Obert *if you see a rich Nigerian youngster tell me I will use my body to get some money for my kids*".

In several African countries, prior to colonialism, establishing a business from the ground up posed significant challenges. For members of royal families, it may have been easier to achieve wealth, as many are usually born into affluence. They would not require education or substantial capital to start a business from scratch, as their family's wealth would cover all the necessary expenses. This is another reason these inexperienced kids treat the rest of society poorly. They are the same ones who will kill, steal, promote immoral behavior, and eventually contribute to economic collapse around the world. Go on social media, you'll see them flaunting their wealth there. They don't understand what investments mean. Even if some of them do, you'll find that only a handful truly invest in Corporate Social Investment, the kind that becomes a country's real investment (CSI). They undermine the real value of community upliftment. If someone educated would say that it only shows they have a shallow mindset.

Unfortunately, for some of us, the journey is fraught with challenges. While writing this book, I was constantly concerned about securing the funds needed for publication, a process that took four years before 'Life Time Runs' could be published.

I think we need to recognize that no wealthy individual has ever reached their status without assistance. This support may have come from family, friends, or even less savory means. Ultimately, they managed to secure the necessary help to achieve their goals. Don't hesitate to seek assistance from your family, friends, or even the government. You are not alone. If I hadn't asserted myself and raised my voice to be heard while on the street, I wouldn't be where I am today. For the South African government to support my upbringing, they undoubtedly made informed decisions. I still struggle to understand how the social workers found room in their hearts to forgive us, despite the challenges we put them through as orphaned children.

What I'm saying doesn't mean I've 'made it' in life, it means I was once seen as less than human. Today, I feel like a human among others because of those who came to my rescue. I'm saying that speaking up and honestly admitting you need help can make a real difference.

Collective circle:

Whether you're employed, unemployed, a part-time worker, a student, a dropout, a graduate, a business owner, a government, or community leader, or even homeless, you likely to fall into one or more of these categories. Your placement is not a matter of chance or a curse; it may be the result of past decisions or influenced by historical events.

If you were in South Africa in 1995, trust me, you had ample opportunity to register a business. It wasn't necessary to start it immediately, but any concept you had in mind should have been registered in your name by then. I'm not denying that you endured oppression, or suggesting you sided with the oppressor, but this was the dawn of freedom, a moment that offered far greater opportunities to carve out a meaningful niche.

No matter the path you choose, there's always a hierarchy to navigate. Whether you own a business, secure government contracts, or operate within global governance, there's invariably someone aiding or impeding your progress. This dynamic is particularly pronounced in

Africa, where local policies often await international legal precedents before adapting to attract and accommodate foreign investments that drive societal development. We should not allow ourselves to be controlled by other nations. We should appreciate that we are living in the post-colonial era and strive to build a stronger economy than the one left by our colonial predecessors. We should proactively dismantle systems inherited from colonial times and strategically manage our resource supply chains to prevent foreign investments from exploiting vulnerabilities.

We need to learn from Ibrahim Traoré, not because he leads from inherited notions of African ancestry, but because he draws wisdom from the colonial legacy and the resources entrusted to us by our ancestors. I like the strategy he's using to reclaim his country — starting with his own people instead of continuing to feed the French empires. Now look: every social media platform is saying that France is ungovernable, and its government is collapsing. My question is, will they look for a scapegoat? And will any other African country fall for it after seeing how Burkina Faso has finally woken up? What's significant is that we now stand behind a leader whose support echoes the legacy of Soviet-era solidarity.

Even if he refuses to place a hierarchy in charge on his behalf, he must remember this: those who support him will one day remind him who truly holds the power. They want something in return. Nothing, absolutely nothing comes for free.

Leakages.

Too many lead with blinded eyes,

Afraid of truth where challenge lies.

Yet fire refines, it does not burn

It teaches those who dare to learn.

Without the spark, the mirror fades,

And discipline walks hollowed ways.

A loyal march, but off the track

No wisdom guiding what we lack.

Economic leakages are a critical issue that must be addressed, though the complexity of the task is undeniable. The Black Market represents our most substantial leakage, and it's essential to tackle this challenge before delving into the broader regulatory landscape.

The government holds the reins of regulation. Recognizing our vulnerabilities is the first step. Unfortunately, the crux of the problem lies in the absence of true leadership; instead, we are beset by individuals driven solely by financial gain. A common pitfall for those in leadership positions is prioritizing personal wealth accumulation over national stewardship. A tactic often observed is the strategic reshuffling of the cabinet post-election, enabling the appointment of allies who will support systems beneficial to their interests.

Our leaders often avoid appointing individuals who might question their decisions, especially those that favor personal benefit over societal progress. Too often, African leaders forget that true growth emerges from overcoming challenges. Without opposition or critique, there's little room for reflection and discipline, unchecked, can become devotion to the wrong path. Some chase dictatorship as if it were a goldmine; others stay silent, fearing their best ideas will be ignored. In the end, they believe the president's word is law, when in fact they are wrong.

Africa is at a crossroads, facing the prospect of ceding more land to foreign investors than to its own people. It is imperative that African leaders prioritize land distribution to their citizens to prevent this imbalance. As remote areas become urbanized, the rush for pristine natural landscapes for economic development will intensify. Now is the time for proactive measures to ensure that the continent's lands remain in the hands of its people.

This must not be pursued through violence, but through professionalism and integrity. We are people of compassion, and our actions should reflect fairness and respect for the essence of humanity. As stated in the first chapter, we carry the most valuable DNA. This is why the world is conflicted unsure whether to eliminate the people who hold the DNA they need, or the rivals who are also pursuing that same DNA and its resources.

Julius Malema said: In 1994, our priorities were misaligned. What we needed most wasn't RDP housing, it was quality, free education. With the right foundation, we would've been building our own homes by now, not waiting in line for government issued ones. Empowered minds create empowered communities. Instead, we've become dependent on social grants, when we could've been architects of our own prosperity. But my question is this: as a Zimbabwean, and with my country of birth once globally recognized for its education, where are Zimbabwe's most educated voices when Malema says all that? Because Zimbabwe is undoubtedly one of the most educated societies in the world, is dictatorship the only force preventing our great minds from improving the nation's economy?

18 August 2025 SA Under Cover Investigation TikTok video: A woman in the video said: there is no truly democratic president in the world. Democracy, she argued, is an illusion, it cannot exist within a capitalist, imperialist, patriarchal, and religious system. The reason is simple: all of these systems are driven by the pursuit of money. Remove religion, capitalism, and imperialism, and what remains is socialism. In her view, that would mean the end of government and the beginning of freedom. For her, the true definition of democracy is freedom.

I've spent years trying to grasp the true meaning of freedom. I still remember a close friend in high school saying, "Freedom is just a state of mind." For the past nine years, I've been quietly wrestling with what he meant. Being a slow learner is both a gift and a burden, it forces me to reflect deeply, but often at a pace the world doesn't wait for. Is this why so many of our communities' struggle to find work or start businesses? Because investors don't have the time or the patience to invest in teaching someone who's still learning? Are they only chasing money through time management shaped by the Western calendar? Now that Eastern culture seems to be taking hold across the continent, what will they do? Will they adapt to the changes, or will they return to their countries and seek another place to feed their colonial ego? I'm not saying this simply because we are the heart of the world. I'm saying it so that we can contribute to positive change, so that we can stop the evil plans of those who have lost their cultures and now roam the world, destroying nature and eroding the remaining cultures across the globe.

The second speech made by a Chinese man: SA Undercover Investigation TitTok profile: Africa must chart its own course free from external imposition. It is deeply troubling that decisions about our continent's future are debated behind closed doors by powers who ask whether we should align with China or the United States. That binary choice is not only misguided, but also a disservice to Africa's sovereignty. The African continent must be powered by African intellect, African innovation, and African vision. The more the outside world interferes, the more it extracts from the richest and most resourceful continent on Earth.

14 August 2025

But even in exile,

the pen remains my weapon

and truth, my shield

Mandy Zvidza; gave a public speech, thanking the Zimbabwean society for their support of the national police service. But he also voiced strong criticism of ZANU-PF. By August 19th, I saw a post on TikTok claiming he had been arrested. In Zimbabwe, challenging the ruling party in public is not tolerated. This incident is yet another reminder that the government resists scrutiny and fears exposure, especially when it comes to the systemic failures that continue to cripple the economy. They do not want to be seen as flawed leaders. If I were in Zimbabwe, I doubt I would have been allowed to write this book. The truth I've put to paper could have cost me my freedom, or even my life.

The irony is that these failed systems are colonial at their core. Even when a system is built in Africa, we know it was designed beyond our shores, which means it remains colonial. Nothing changes.

There's something deeply unsettling about Zimbabwe. The government has been systematically silencing voices, whether they belong to opposition party members, journalists, or even police officers who dare to challenge the corrupt machinery. And now I find myself asking: if they discover I'm the author of this book, will they try to silence me too? Will they come after me for exposing these truths? I also wonder, how connected is the outside world to what's unfolding within Zimbabwe's borders? Are they complicit, indifferent, or simply unaware? What I'm sharing reflects only my personal view, formed from the way I see and understand things.

We must understand this: the president of a country is not truly in charge. In reality, powerful hidden networks often referred to as secret societies are believed to influence and direct the course of events from behind the scenes. The leaders of many nations today are little more than symbolic figures, placed in office to represent the will of these unseen powers. These groups dictate how countries are run, shaping policies and decisions to maintain control over the masses.

Consider this claim: the president of the United States is said to have areas within the White House that remain off-limits to him. In those shadows, it is not the elected leader who decides what to do, what not to do, and when to act but rather an unseen authority that issues the real instructions.

Before the emergence of COVID-19 in 2020, Bill Gates highlighted concerns about Africa's population growth. His financial influence as an investor can impact decision-making in parts of Africa. For instances, while innovative solutions for economic development from individuals may go unnoticed, a prominent figure like Bill Gates can easily garner attention. Leaders may be inclined to support his views, anticipating potential benefits from investment projects. During that time, it was reported that they acquired significant amounts of land both in Africa and in certain Western countries. However, it's crucial to note that Bill Gates advocates for health and education investments as solutions to population challenges, not depopulation. Prioritizing solutions that contribute to societal development is far more important than relentlessly pursuing wealth and status. That's what we are missing.

Figures like Bill Gates can only gain advantage and invest in ways that exploit Africans through the cracks left by weak leadership. It is our own leaders, driven by personal gain who permit such influence, often failing to look beyond the polished promises and confront the deeper intentions behind wealth and power.

The uncanny ability.

Sometimes, we may mislead others by claiming we possess certain skills when in reality, we do not. This can stem from a desperate desire for recognition, leading us to overstate our capabilities. Such individuals are often labeled as 'know-it-alls.' True leaders tend to avoid associating with those who claim to know everything. Being perceived as a 'know-it-all' can result in others distancing themselves, as it inadvertently suggests a lack of genuine knowledge.

These are not myths; I speak from personal experience. Perhaps that could be another topic of discussion, I found my path to writing through lived experience, yet in the eyes of the so-called true writer, I remain an outsider, unwritten and unrecognized. Some writers and so-called senior citizens often say I lack the skill to write. Yet I believe a true writer can recognize when someone has the potential to become the person they aspire to be. That's why I always say: don't just give advice, give the resources. Equip the person or organization to lay the first foundation and let them grow by learning through failures.

Consider this perspective: Many leaders in our nations have been shaped by the legacy of colonialism. Some may lack formal education, yet their primary concern seems to be the accumulation of personal wealth. While I don't want to imply that they lack financial education, it's essential for leaders to be well-qualified in areas beyond just accumulating wealth in their bank accounts.

The pursuit of economic development within their own countries, or Africa as a whole, often takes a backseat. A friend of mine ask this question "*How can you be a multimillionaire, Billionaire or a Trillionaire, in a poor country or region?*" This self-serving behavior is sometimes justified under the guise of fighting for liberation. As individuals who have grown up in the post-independence era, we cannot claim to possess all the knowledge and skills necessary to support our views on community development.

If our leaders continue to appoint those whose relevance was tied to the colonial struggle, the world will keep viewing Africa as dependent forever in need of aid. It's time we shift our focus. We must prioritize those with the education, vision, and capacity to invest meaningfully in African soil. This is our home, our wealth, our future. Let us invest in it, not just export it. The outside world thrives on our division, because a united Africa is far too powerful to be ignored. That is the truth they fear (*For when Africa stands tall, the world must look up*).

We need an education system that cultivates strength of character, one that empowers us to resist the temptations of greed and weakness, especially in the stewardship of national finances. I speak not from academic privilege, but from lived experience. With only a primary education, I may not offer technical solutions, but I understand this: true leadership demands selflessness. It requires the wisdom to balance a nation's needs with its wants, much like parents who must nurture both discipline and dreams in their children.

Leaders must abandon the pursuit of Western celebrity endorsements and quick funding schemes, which often open the door to corruption. The obsession with personal wealth must end. Our leaders must return their focus to the people, to the communities they vowed to serve. We must not allow foreign entities to outperform us in the very culture that belongs to us. The world has already taken much from Africa, and now is the time to reclaim what is ours, our cultures and our resources. It was the rest of the world that was meant to seek from us, not the other way around.

Equatorial Guinea.

It's time we rewrite the story

one led by truth, not tyranny

This country stands as one of the clearest examples of African wealth being siphoned off to the West. Its leader is a figure who revels in luxury while silencing opposition. Reports suggest he owns properties across the globe, some of which he's never even set foot in. His sons too, benefit from the same wealth, drawn from oil exports that rightfully belong to the people. When ignorance persists, especially when passed down through generations, it becomes a legacy of destruction. And that legacy is costing us our future.

The legacy of a leader even when carried forward by their sons, must be critically examined to ensure it serves the greater good. The fear surrounding Mugabe wasn't just about his biological heirs; it was about the endurance of his ideology. From an educational standpoint, they feared his legacy might evolve beyond dictatorship into something more deeply entrenched. And today, that legacy lives on, not through his children, but through those bound to him by blood and loyalty. Ironically, many of them claim to have fought for freedom. Could this be the foremost reason, and the weapon most often employed, to subjugate the weak and those deprived of education.

Yet in today's reality, those who dare to improve the economy or challenge the ruling party face imprisonment or worse. True freedom of speech exists only beyond the borders. Within the country, silence is survival. To speak against ZW or any authoritarian regime is to risk everything in Africa. I am unfortunately not going to say much about the country because I do not have reliable information.

Most outstanding lifestyles:

In the quiet war

between greens and grease,

each choice whispers

a future into the bones

some woven in strength,

others frayed by time.

Healthy lifestyle

The impact of healthy and unhealthy people on the economy can vary significantly based on regional factors such as healthcare infrastructure, education, social support, and economic development. The demand for products and services varies based on their effectiveness and costs. These factors play a crucial role in shaping consumer preferences and overall market dynamics.

These products don't come cheap, and that leads manufacturers to import large quantities from the secondary sector. If these products and services are not imported, they may be manufactured rapidly, often using harmful chemicals, which can contribute to environmental issues. The products often contribute to an unhealthy lifestyle, and it's evident that they were designed for those who want to appear wealthy, even if they are economically disadvantaged. During production, a significant number of harmful gases is released into the air. Plastic bottles and other plastic products are major contributors to pollution in reliable sources. I commend large corporations that actively support and collaborate with recycling startups. Their commitment to fostering innovation in sustainable practices contributes significantly to environmental conservation and circular economy initiatives.

What's even more troubling is that this waste is routinely dumped in regions where Africa's most vulnerable communities live, turning poverty into a dumping ground for global negligence. Whether this is a deliberate oversight or a lack of resources, these regions often struggle with inadequate cleaning and a lack of access to a clean environment. The government faces major challenges in addressing these issues due to constraints related to time and finances. When governments are consistently challenged or feel overwhelmed, they frequently resort to seeking fast and inexpensive labor for constructing infrastructure. This type of solution or direction, when taken by governments, can sometimes harm economic growth. The government has the last say in public but not behind closed doors.

We need to encourage the establishment of more recycling companies, and the government should actively assist new cleaning and recycling start-ups. Securing funding for a business that's just starting can be particularly challenging compared to supporting an already established business. Prioritizing support for smaller companies can yield rewards beyond our imagination. Their agility, innovation, and potential for growth contribute significantly to economic ecosystems and foster resilience in the business landscape.

UNHEALTHY:

The well-being of communities has a substantial impact on the economy, and unhealthy communities are more likely to negatively affect economic outcomes. While we often play a crucial role in supporting the formal economy, it's essential to recognize that, at times, we inadvertently contribute to the black market. Our actions can have far-reaching consequences, impacting economic stability and societal well-being in ways that may not always align with legal or ethical norms.

Unhealthy, we demand households goods and services and we are always in pursuit. We are always striving to obtain everything at a lower cost. I understand that affordability is a significant concern for many people but in Africa the desire for free access to services and resources is beyond manageable at times. Urging governments to provide free education reflects a commitment to social justice and equal opportunities. The government faces constraints as it cannot provide for everyone simultaneously.

Inflation is a global economic phenomenon that affects countries around the world. When inflation occurs in one part of the world, it can indeed have ripple effects on other regions, especially in African countries. Most African countries that rely on export may face challenges. These are indeed challenging times for African countries, as they grapple with the impact of global inflation. When quality products are primarily exported, local communities usually struggle to access nutritious food. This can result in food insecurity, malnutrition, and unhealthy lifestyles. Crime rate increases during these times and it is still going to, due to the population growth.

Obert "As a boy, my belly often swelled mysteriously, even on days when I had eaten nothing at all. It puzzled me then, a silent riddle in my own body. Only years later, growing up in South Africa, did I come to understand the name for what I had carried kwashiorkor. In that moment, a piece of my childhood unveiled itself, and the story of my hunger finally spoke its truth" If you search the words *"hunger in Africa"* online, you'll likely see images of small Black boys, shirtless, barefoot, their stomachs distended, wearing only shorts, their skin coated in a film of white dust. This was simply another image I was trying to paint for you, showing how a child with kwashiorkor might appear. The reference pictures are ones I found online, as I do not have any of my own.

The relationship between the developed world and Africa's developing nations is far more complex than it appears. While the narrative of wealthy countries generously providing aid is widely propagated, that aid is rarely without conditions. Cycles of resource extraction, unequal trade, and relentless debt repayment deepen existing inequalities. In the end, many communities are left weakened, either malnourished or slowly poisoned by food laced with harmful chemicals. This is done so that the same developed world businesses can return, repeatedly, to strip Africa of its natural resources, until nothing remains. Do they truly wish to erase Africa to this extent? Must they be so ruthless, all in the name of living in luxury? Is this so they can forever parade themselves as the pinnacle of 'first world' advancement.

Aid should be provided without imposing harmful conditions that perpetuate dependency. Let's implement a system for more equitable relationships between nations. We should perpetuate the awareness that we Africans have everything. The aid we receive is often made from our very own exported raw materials, returning to us as little more than breadcrumbs. We do not need this. We have the capacity to manufacture the very same products now sent to us under the banner of aid, so why must they be shipped back to us? Why must we sign endless documents to access foreign financial assistance? And when disease outbreaks occur elsewhere, we must protect our people, by halting the influx of foreign nationals into the continent until they can prove they are healthy enough to be here.

Alcohol:

Does Alcohol and drugs bring more income in most parts of the African continent?

In 2000, alcohol was the most important risk to health in low and middle income countries in Africa. However, it's important to note that while alcohol consumption is prevalent, it doesn't necessarily translate directly into income for the continent. A cross-sectional study of middle-aged adults in four sub-Saharan African countries (Ghana, Kenya, South Africa, and Burkina Faso) found that more men (68.4%) than women (33.3%) were current substance users.

The saying that the richest place is not necessarily one of those cities with highly shiny skyscrapers or the highest GDPs in the world is thought-provoking. The true wealth of a place often lies beyond material possessions and financial metrics. The richest place is the human mind, where knowledge, creativity, and ideas reside. That's the Intellectual capital, innovation, and the ability to generate new solutions are valuable.

Substance abuse is just another waste of these untapped potential and unexplored ideas within individuals. Instead of leaving our potential buried and unrealized. Let's contribute to the world, create meaningful connections, and leave a legacy that enriches lives. We cannot keep enriching graveyards by destroying innocent lives through the scourge of drugs and alcohol.

I have consumed alcohol and smoked in my life, so I believe I have no right to judge someone else. We all possess our own imperfections. During my high school years, I experienced the most fragile times when peer pressure was at its utmost as a youth. If I had been raised in Zimbabwe, there would have been less chances for me to indulge in such bad habits such as consuming alcohol and substance abuse. I had no intention of drinking alcohol or smoking, yet I would've still found myself becoming a father in my teenage years, a generational pattern common across many African countries. In much of Zimbabwe, this way of life is still quietly endorsed and upheld.

In South Africa, the alcohol and drug trade often generates more income than traditional employment. Vigilance is crucial, especially when dealing with businesses that operate in less conventional ways. While some may appear successful initially, it's essential to recognize that those who engage in such practices often face challenges and may not sustain their operations over the long term. For those residing in townships, there is a higher likelihood of becoming involved in one of these two industries. The unfortunate reality is that once individuals find themselves in such circumstances, they are more likely to end up on the streets.

The economic realities and social conditions in these areas can significantly influence individuals' choices and opportunities. It is indeed a sobering reality that many drug addicts and alcoholics in South Africa often lack access to proper rehabilitation services. This situation can have a negative impact on both the individuals themselves and the broader economy. When individuals do not receive adequate treatment, it perpetuates a cycle of addiction, affecting their ability to participate in the workforce and contribute positively to society. Since my parents can't work, I have no food to eat, no school fees, and I can no longer live in safe, secure environments. That leaves me with impossible choices steal, work at a young age, or even turn to illegal activities like drug trafficking or selling alcohol without a license. And just like that, my circumstances have already pushed me into actions that harm the economy itself.

We must find solutions to this growing problem, it seems to be worsening by the day. When I was growing up, the idea of seeing someone injecting themselves with drugs on the streets was unheard of. Now, it happens openly, in plain sight. The situation is spiraling out of control. There are areas and suburbs where police can no longer enter to do their jobs. They stay away because innocent civilians could be caught in the crossfire when drug dealers resist law enforcement. It may also be that some of their own family members live there. They're not willing to risk their own lives.

Personal experience.

Du Noon, a township in Cape Town, South Africa, is known for its unique lifestyle and challenges. On the afternoon of December 13,

2023, I found myself sitting in my room, a space that was less than a meter in width and only 4 meters in length. The cramped living conditions in Du Noon are a stark reminder of the socio-economic disparities faced by many residents.

Despite the limitations, the resilience and spirit of the community continue to shine through, as they navigate their daily lives in this vibrant and dynamic township. Looking at my furniture; that's a single bed, one quarter meter shelf, empty Bosch fridge, 2 burner gas stoves. My eyes were blurred with tears, while introspection weighed heavily up on me. I question myself what am I doing with my life? Why am I torturing myself. I have a nine to five job and yet I am not fixing my life. While some people have wives and businesses, they manage to take care of their families with the little they have. I was inspired to see how resourceful individuals can provide for their loved ones even in challenging circumstances while on the other side I seemed to wasting opportunities and crucial resources I had. When I first arrived in the area, I was still new and unfamiliar with everything. While I was busy writing a book, I also printed a few copies of my business concepts. Despite visiting numerous businesses and government branches, all of them denied my request to fulfill my dreams based on the ideas I had. To be quite frankly I felt disheartened despite how persistent I was while facing such challenges. It felt strange when I approached other African individuals to present my concepts, only to be met with rejection. The rejection was beyond the way I expected it to be, since the individuals denied me being in contact with his superiors recalled racism treatment I have experienced several times in my life.

We block each other from accessing opportunities due to financial reasons and long term networks. If you had to measure my net worth at that moment it was obviously zero which made the period even worse.

When I take a moment to reflect, I realize that I haven't made significant progress yet. I felt frustrated that I had nothing. I am aware that my ideas and books have the potential to generate billions. Although I recognized their potential value, I also knew that as long as they remained in my hands, they would remain valueless until they were in the right hands. The right hands, in this context, are often associated with lighter skin color. Skin color can sometimes speak as loudly as money does. In certain fields, when you don't get the opportunity you wanted, you're often told to be accompanied by a white and so do you then succeed.

That's reminds me of a video of Julius Malema saying he appreciates his white bodyguard, because whenever they travel or enter a business, doors are opened for him more readily than when he is accompanied by his black bodyguards. That shows how, in some places, white skin is the most feared. In certain areas, simply speaking English is seen as a sign of wealth. Should it be a shame when they discover you are just as poor as they are, though the language may be different. When I returned to Zimbabwe in 2019 to register, only a few people wanted to be my friends, yet some of them are still trying to reach me today. The attention became so overwhelming that I had to delete some of my social media accounts. In the end, having too many of them in my life has left me with nothing. Today, I'm the one forced to beg just to survive, while those who took from me are nowhere to be found.

African wealth is not lost; we possess it, and it will remain with us. The challenge with our wealth lies in the fact that it often exists only in our minds, without tangible manifestation. We are squandering more precious time while we are blaming the disadvantages caused by the history. Yes, I admit that happened though I was not yet born but since I am also lazy, I do also feel guilty at times. Instead of sitting and seeking the most profound excuses and blasphemy to the history let's stand up and explore available opportunities. Despite the disadvantages, I believe there are more opportunities available.

Zimbabweans in SA.

Zimbabweans are well-known for their resourcefulness and ability to find quick survival solutions. They often find quick survival solutions to meet their immediate needs while also learning from their mistakes over time to embrace growth within their allocated valued resources. It's unfortunate that some Zimbabweans in South Africa face challenges due to negative perceptions. While it's true that there have been tensions related to job theft by Zimbabweans. But if you look closely at the history, you'll see this has been an ongoing war, a conflict between tribes. I can be honest with you, the two countries do carry the same DNA. While South Africa has 11 official languages, Zimbabwe now has 16 of which roughly 75% are also South African languages. They share the same culture and history, yet still they fight, as if the echoes of their ancestors' unity are no match for the shadows of pride, power, and wounds left to fester.

How does one take another person's job, by going to their boss to have them fired, or by resorting to blackmail?

And how does their presence in South Africa truly impact the economy?

Is the Zimbabwean government driving its own citizens out?

Has the post-colonial regime turned its revenge on innocent people instead?

If so, why are we allowing it to happen?

Is there a white supremacy network somewhere in Europe still investing in events designed to divide us? Do they work to a deadline, or a target number of lives to erase?

These individuals can never orchestrate an event that will result in anyone job dismissal. They can never blackmail you to lose your job. I must admit that they are willing to take any type of job just to survive. Being Zimbabwean means taking life seriously; your work ethics have to be on point. We are just like any other nationalities, where we have diverse experiences and approaches to life. Zimbabwean knowledge and strong work ethics can sometimes make you appear inadequate in front of your long-term employer. We rarely complain; if we feel we are earning less, we will either seek another job or start a side hustle. This such tremendous resilience and adaptability of being a Zimbabwean during the economic challenges allows us to remain immune to long term struggles. We have a reputation for being resourceful and entrepreneurial. Despite facing economic challenges, many Zimbabweans demonstrate resilience and adaptability by exploring various business opportunities. Jobs theft in South Africa has contributed to tensions and safety concerns. That gets accompanied by mistakes done by any other foreignationals while 80% of the blame is shifted to Zimbabwean nationals.

As a Zimbabwean, navigating life in a certain region of South Africa can indeed be challenging due to various factors. The historical tensions between the Shona and Ndebele & Zulu ethnic groups in Zimbabwe have indeed contributed to complex dynamics, including xenophobia. You need to understand that Ndebele has a sibling which is called a Zulu. So does these ethnic groups carry the same hatred of their brothers and sisters. I have been informed that some of them indeed engage in practices that prevent many Shona people from accumulating wealth. I was told that I am mix of Malawian and Shona and I wonder if they will bring that to celebration since I embody rich cultural tapestry or will they harbor biases and or act discriminative?

You might wonder where this affects the economy, well in each and every place that is overpopulated does not have the best hygiene. There are not enough public services for everyone.

Of course, I don't blame locals for wanting to protect their economy from being undermined by other nationalities. What I cannot support, however, is unfair treatment, especially towards highly educated individuals. Julius Malema and Mmusi Maimane were not calling for undocumented migration without order; they were advocating for documenting those without papers and opening the borders. Their message was widely misunderstood. For some, particularly the less informed, this idea translates into higher taxes. But for the educated, it represents African unity and the vision of building one strong, shared economy.

The former South African Minister of Home Affairs, Chief Inkosi Mangosuthu Buthelezi, cited three reasons for the increase of Zimbabweans. He stated that South Africa, compared to other African countries, was economically advanced, that South African employers were more willing to hire foreigners including vulnerable and undocumented (i.e., illegal) immigrants, and that Southern African countries were historically and economically connected to South Africa which, in and of itself, influenced migration patterns.

Individual economic challenges such as poverty, unemployment, and homelessness, may not however be the only factors contributing to international relocation. Institutional and structural factors such as civil unrest, wars, and political instability may also serve as motivators of immigration to other countries and have yielded almost 15 million refugees and 22 million internally displaced persons throughout the world. Furthermore, the consequences of civil unrest and political violence commonly result in individuals being deprived of basic necessities such as food and water and being exposed to severe and threatening experiences such as tortureand/or imprisonment.

I want dive into the paragraph cited above by the former DHA minister, with a particular emphasis on the economic challenges. Despite the fact that some of the newly arriving ZW use the same old reason, this particular fact remains unchanged. The challenges highlighted by the former DHA minister continue to persist in Zimbabwe today. Thus significantly affects South Africa's economy. The South African government do provide services for everyone in the country, despite these external pressures. As the population continues to increase, these services become scarce. While the South African economy may be advanced compared to other African countries, it is still in the developing phase. The increasing population leads to economic strain. I do applaud the South African government they way they do handle despite the continuous political dramas repeated to entertain some of us that vouch for fixed and neglected growth mindset. I also find funny that the SA politicians knew the skills to keep their people entertained.

Could that be another marketing strategy? You're probably wondering how. Well, they keep you entertained while, behind the scenes, someone is working on the very opportunities you're letting slip away distracted by the spectacle of parliamentary debates. The next day, you see someone opening a business nearby and you're stunned, wondering how they pulled it off. The truth? They built it while you were watching TV, out partying, marching in protests, or busy chasing away so-called 'illegal' foreigners.

The arguments in parliamentary houses may keep smiles on the faces of citizens, yet they quietly drain the economy. I tell you, these same entertained citizens are being robbed in the process. These are the same citizens who will turn around and tell a foreigner to leave. These are the times when what truly matters is ignored, only to receive attention later, when decisive action could have changed the outcome. Then you only spoke of what should have been done when it was far too late to do it.

Now, you might wonder how this relates to Africa's lost wealth.? Well, our leaders lack the actual leaderships skills. The role of leadership is crucial in addressing the challenges faced by countries like Zimbabwe. When leaders prioritize transparency, accountability, and effective governance, it can significantly impact the allocation of resources and the development of essential infrastructure such as roads, schools, housing, agriculture and hospitals and technology advancement. I am aware and I do recognize that the issue extends beyond Zimbabwe alone. In some cases, a government official might ask you to give them money to fix the country, but unfortunately, they often prioritize their own personal interests over the nation's welfare.

Engaging in protests can be counterproductive and potentially dangerous, leading to unintended consequences. While working in Franchhoek, I often advised my colleagues against protesting, as dissatisfaction with work and salaries does not always warrant such actions. Often, those you protest against may already possess everything they need, and being seen throwing stones at passing vehicles could result in job loss. One of the reasons Zimbabweans are compelled to leave their country is the realization that protesting for change often proves futile. They are aware that it's only a matter of time before they are pursued and potentially executed for organizing such events. Now you know that we are kept in poverty in purpose. This has been beyond our control, but it becomes our fault if we fail to keep demanding economic improvement.

Challenged by young and fresh bloods.

In Zimbabwe plutocrats would use Mugabe's regime as a means to block out competitors. When you're the wealthiest person in the poorest country, while there are opportunities for anyone to challenge you, it's only natural that you would feel uncomfortable with that. When such situations arise, individuals may find themselves compelled to seek help, often resorting to corrupting the authorities. Within authoritarian regimes, there are often existing financial vampires who exploit their positions for personal gain. Such strategies have been integral to the economic landscape for many years, often shaping market dynamics and influencing outcomes. We are born into this world, and inevitably, we will depart from it, leaving behind the legacy of our existence.

Do not involve Zimbabweans in drug trafficking.

Gold flows where the sickness stays,

White coats and shadows count their praise.

Quick thrills trade in borrowed breath,

Both chasing profit, not the depth.

One sells the cure, one sells the high,

Each feeds the hunger money can't buy.

Anyone can do anything that has something to do with stealing labor in a foreign country but you can never include ZW in drug trafficking. In 100 there's definitely at least 5% of Zimbabweans that maybe probably involved in drugs trafficking.

Pharmaceuticals are a lucrative industry, and their financial performance often outpaces that of many other businesses. Instant gratification plays a significant role in the appeal of drug trafficking and other elicit activities. When someone is not involved in drug trafficking and does not actively generate income through illegal means, we can only considered them as a consumer. When traveling to a foreign land it's essential to be mindful of local customs and contribute positively to the community. I'm not accusing you of using drugs; I'm simply pointing out facts you should be mindful of facts that directly affect you as a responsible individual.

This means benefiting from the country without giving anything back, unless you hold a professional qualification, such as being a doctor, pilot, plumber, soldier, or other skilled professional, your contribution is minimal. Zimbabweans rarely engage in illegal activities like drug trafficking, yet they often shoulder more than their fair share in informal labor. When it comes to work, we often take on more than is expected without even realizing it. Now, consider a Nigerian equipped with IT and pharmaceutical skills. This makes it easier for them to analyze the pharmaceutical drug supply chain, identifying high-demand locations and setting competitive prices. In such circumstances, no one is interested in a Zimbabwean who cannot generate quick profits when compared to a drug trafficker.

I've observed that people of color are disproportionately affected by various forms of exploitation and disadvantage. They are experiencing a higher rate of drug-related deaths compared to other racial groups. This is not a government recorded but my perspective aiming to contribute to a broader understanding of the world. I grew up in a multicultural environment with diverse personalities, that provided me with unique and enriching experience. It has been suggested that they are often targeted and exploited, and unfortunately, they may also be more vulnerable to drug-related issues, especially young males. Not only are they exploited by their fellow citizens, but other African foreigners also use their females to obtain legal documents through fake marriages and other related relationships. The question of how individuals end up in certain circumstances, such as homelessness or addiction, is multifaceted and influenced by various factors. Well the answer is simple; drugs and alcohol. In the children's home, I was told that in a colored communities becoming a man is associated with becoming a gangster and spending time in prison while being strengthen by other experienced gang leaders through same gender sexual intercourse, speaking gang language in a particular style and being tattooed based on the ranking within the gangs levels. The ranking exists only in their minds, not in reality, yet it grips them until their last breath. In gang life, there's no easy exit; try to leave and you're marked for death. The only way out is when you stop breathing.

While their women may find themselves begging on the streets, being another relationship, due to challenging circumstances, their husbands are serving in prison for whatever crime may have been. Sometimes you may discover that some of the were falsely accused of crimes they did not commit and were unfortunate enough to end up in prison. Some may have deliberately committed these crimes just to experience prison life, claiming it made them a "real man," only to regret it in their old age. Some remain in the gang because the outside world would swallow them whole; step by step, they lose the skills to live beyond its walls. They observe the news later in the evening, where the government promises employment opportunities in exchange for their votes.

I am aware that some women engage in **prostitution**. However, like everyone else, they often ensure that their activities take place far away from home. While I don't endorse or support prostitution, I recognize that some individuals may turn to it as a means to provide for their families. As a community, we have failed to provide them with adequate education and employment opportunities, which can lead individuals to make difficult choices. Despite that being an excuse used by some, it may be a behavior or mindset passed down within their households. That makes it difficult for them to explore alternative ways of living.

Let's discuss how this situation affects Africa's lost wealth. The challenges faced by women in various African communities, including economic hardships and limited opportunities, have broader implications for the continent's overall wealth and development. When individuals experience such hardships, it indeed feels like they are being robbed of their life and wealth. The cycle of poverty, lack of opportunities, and economic struggles can perpetuate this loss. We perpetuate white collar crime of theft and that is a significant concern. If we do not solve white collar crime then, there's no way the African economy can develop. I believe white-collar crime laid the foundation for the black market, its hidden deals and quiet corruption paving the way for underground economies to thrive. It may seem as though white-collar crime is the only path to financial access, perhaps that's simply how it appears. We must not allow that, as it's often how the problem begins. Now you understand why law enforcement is often deployed to arrest those on the streets, they make their activities too visible. Meanwhile, behind closed doors, in the home of a man in flip-flops, khaki shorts, a white-striped polo shirt, and sunglasses, lounging by the pool and sun-gazing as he runs his fingers through his blond hair, the real deals are being made, phone pressed to his ear.

Poor infrastructure negatively affect:

The type of transport we use has a significant influence on our economy. The private trucking industry's efficiency does not adequately align with supply and demand dynamics in Africa. In addition to examining the private sector, it is essential to consider the state of our national roads when evaluating transportation efficiency.

The prevalence of accidents and armed robberies has become a significant concern. The transport industry in Africa heavily relies on and operates within Western systems. Our economy can only grow sustainably if we prioritize improving the quality of our internal supply rather than solely relying on exporting primary materials. We must utilize resources within our own borders, efficiently and innovatively, to add value and contribute to long-term economic stability and resilience. While exporting is often seen as an indicator of a thriving economy, unfortunately for Africa, it can sometimes mean that foreign markets overpower our economy, treating it as prey. Promoting and supporting the local market is crucial, especially in post-colonial contexts, as it allows for economic independence and resilience against foreign market dominance.

Despite lacking a robust secondary sector, we strive to create a positive impression by exporting substantial quantities of high-quality goods, with the hope of enhancing market value. In my assessment, relying solely on the strategy of exporting large quantities of goods without a robust secondary sector poses a significant risk to our own economy. To strengthen our economy, it is imperative that we prioritize acquiring high-quality education to build a robust secondary sector. By focusing on producing finished products rather than solely relying on primary resources, we can enhance our competitiveness in the global market. This does not deny that producing tertiary goods within our own borders may be more costly, but it could position our markets to dominate the global stage.

Between 1884 and 1914, during the era of European imperialism and after World War II, Africa found itself sitting on vast natural wealth. This are the periods where European powers exerted control over African territories. The extracted and exported, shaping the continent's history and economic dynamics. Bartering trading market was the most valuable platforms for exchanging commodities among Africans. While bartering trading as valuable platform for exchanging commodities among Africans, evolution inevitable in all aspects of life. Associates progress, economic system adapted, and new methods of trading emerged.

While people in the Middle East used camels, Americans relied on American mules, and Asians utilized elephants for transportation, Africans predominantly employed oxen and long-bed chariots, often accompanied by donkeys. A few centuries later, transportation technology has significantly evolved, making it easier to invest in more efficient and advanced transport systems. While many technologically advanced nations have positively influenced and invested in African transportation, these efforts often come with certain conditions and implications.

Technologically advanced nations positively influenced African transportation by introducing modern technologies, better road networks, and improved logistics systems. The problem is that implemented this system based on their cultural and historical context. Instead, today only the imperfections of this system rise than meeting the need of transport in Africa. The current transport status is a very critical concern. It can be deeply disheartening and embarrassing to join a guided tour with tourists on certain road trips in Africa.

What's even worse is that some transport systems overseas are designed in Africa by Africans yet never used on the continent. Instead, foreign investors have long exploited African resources to serve their own interests, offering far less than their true value in return. These are precisely the services we need, yet we are handed the opposite.

Soft grudge.

Joyner Lucas's: *I'm Not Racist* is a provocative, two-perspective track designed to force an unfiltered conversation about race, prejudice, and misunderstanding in America.

Although we are living in the post-apartheid era, our grandparents haven't forgotten the past. They harbor what we might refer to as a **soft grudge**.

Why is it important to state this fact?

Against whom do they harbor this grudge?

Let's consider the complex interplay of love and hate within our parents' relationship as an illustrative example. Consider this scenario where, as a reader; *your parents are no longer together, but both are still alive. This situation can be emotionally complex, as it involves navigating the dynamics of their relationship post separation while recognizing their continued existence and impact on your life. Each time they see their child, they're reminded of their past relationship and the reasons for its end, yet they remain bound together by the love they both share for that child.* The emotions are complex love and resentment intertwined. Similarly, consider the soft grudge held by older individuals who grew up during or were heavily affected by colonialism.

When these individuals encounter young white people, they experience what we call post-fear. It's a mix of emotions intimidation, memories of oppression, and a sense of unease. The colonial legacy has left scars and encountering the descendants of those who perpetuated that legacy can evoke a range of feelings. The soft hatred is a result of historical context, and it shapes interactions even in the post-apartheid era. I am a soft person and PTSD has take its tour at countless times. If you had told me, when I was young, what our ancestors endured, I would have laughed it off. But now, through the lens of my own experiences, through the weight of my family's treatment toward me, I can feel that pain. Each time I think about it, I cry, especially when no one is watching. Sometimes I cry out loud, just to release the pressure inside.

I can almost touch the agony in the stories of those who lived through apartheid: the rope cutting into flesh, days without food, the burden of carrying loads far beyond one's strength, and the endless miles walked under a burning sun with a throat dry as dust.

Since the end of colonization, many African nations have ceased active fighting, yet there are still grudges held by the elders within their hearts. These feelings are a testament to the enduring impact of colonial history. Interestingly, it appears that the colonizers themselves also carry post-fight emotions. Elders remember the injustices, the loss of culture, and the struggle for independence. As someone who grew up in a free world, my experiences are undoubtedly different from those of our forefathers who lived through colonialism. While I sympathize with the post-colonial regime and recognize its impact, my feelings cannot fully mirror theirs.

In our interconnected world, let us recognize our shared humanity beyond the boundaries of skin color. It is very essential to look at someone beyond external appearances and delve into the richness of their personalities. Observe how their lives transform as you positively influence their growth. Education is indeed the most crucial tool for positive transformation in the human race. When we influence growth through education, we empower individuals to reach their full potential and contribute meaningfully to society. Can we draw wisdom from both sides of the stories told by those who endured apartheid and colonialism? Today, many of the people I encounter greet me with a smile and the gift of their advice.

Nelson Mandela: "*education is the most important weapon which we can use to change the world*"

How and what do we educate each other on? It's heartening to see that some individuals who grew up during apartheid have acknowledged the wrongs of the past and are actively working towards positive change. Some apartheid kids have chosen to adopt children who lost their parents due to the oppressive regime. By providing love, care, and a stable home, they contribute to healing the wounds of the past. Some individuals are actively involved in community development projects, building infrastructure, creating job opportunities, and promoting social cohesion. Some engage in dialogue, bridge building, and reconciliation efforts. They understand that forgiveness and understanding are essential for a harmonious society. All this efforts are to change lives to a positive state by recognizing that, we are all humans under the sky.

Thank you for your service.

Visiting old age homes and engaging in conversations with retired individuals provides valuable insights into their experiences. Not all retirees have access to pensions, and examining factors such as nationality and race can reveal disparities. Listening to the types of pensions they receive can also shed light on the challenges they face.

Many of the retired government workers in old age homes have diverse backgrounds, including military personnel, doctors, chefs, and builders. When discussing retirees, we refer to individuals who have tangible evidence of contributing significantly to society. Retirees who have made significant contributions to the country's economy through their government service play a crucial role in shaping the nation's financial well-being. Regardless of whether someone receives official recognition, their contributions, even if unaccredited, can still have a positive impact. While I sympathize with those whose unrecognized efforts may have gone unnoticed, it remains their responsibility to step forward and present their case." Not all retirees receive a substantial pension. Life's inherent unfairness can be disheartening, yet it also presents an opportunity for us to take initiative and strive for positive change.

One of the primary reasons for such disparities in pension amounts is undoubtedly corruption. While some individuals fight back using legal means, others find themselves burdened by taxes that ultimately impact their children. Individuals attempt to address these disparities using traditional human rights mechanisms, although in South Africa, these avenues often yield limited results. In today's world, money often wields significant influence, leading to instances where individuals allow their rights to be violated. Your question will be how does this hinder the economy? Some of the individuals in such circumstances are business owners. Not only do they leave money with the family, but there are also certain laws in place to govern these businesses. Some influential individuals actively influence changes within new business laws to align them with their interests. Some of these influential individuals also wield significant international influence. Nearly every country in the world holds great respect for such veterans. This respect often stems from their service during World Wars or their association with ancient royal families, which is considered deserving honor. Some of these veterans were also part of the most brutal white colonial regimes. It is challenging to express gratitude for military services, especially within the context of colonial regimes.

And yet, as I walk through the city today and meet some of them face to face, they greet me with warmth. They have lessons to share, even as they are near the end of their lives, believing they have little left to give. But somewhere inside me, a voice still says, *"This could be one of those who once killed Black people during apartheid."* And still, I choose compassion.

Lose and leave.

In the dust of stolen lands,

hope walks with heartache.

Babies leave in their mothers' arms,

carried past borders of fear.

The past speaks,

we have lived this before.

Yet still, smiles survive the ruin.

During an intense situation, sometimes you find yourself at a crossroads: fight or flee. In the context of economic strife, the rules of engagement shift. Africans endured the painful loss of what was rightfully theirs, as their lands were forcibly colonized by European powers. In several African countries, the most rural areas remain sparsely populated especially in Zimbabwe, the cycle of hope and heartache repeats itself. Babies are born into a world where their parents, driven by desperation and dreams, make a heart-wrenching choice: to take their precious child away from their homeland, seeking refuge in foreign lands. It's a journey fraught with uncertainty, peril, and the weight of responsibility.

They have no choice, for they know it well, they or their forefathers, have lived it before. What will they feed their child? Will that child have the chance to learn? To grow? Too often, a life begins in the shadow of poverty, with little hope for more.

This thought haunts me. If I had not escaped my family's abuse, I would never have received even a basic education, never mind the ability to write this book. And yet, this is not to say Zimbabwe lacks education. On the contrary, it is among the most educated societies in the world. So what, then, has gone wrong? Again and again, the answer points to leadership. As a white handyman in Cape Town once put it: *"The countries in Africa are being misled, that's all, nothing more."* He spoke with full awareness of his words, a seasoned South African senior with remarkable handyman skills. I respect him deeply, not just for his talent, but for the quiet humility he carries, even when the pressure of a busy day draws out a few choices swear words.

Rich African living abroad.

The lifestyle of some affluent Africans mirrors a survival of the fittest ethos a delicate dance between opulence and competition. The word " **King**" carries a dual nature, a dichotomy that encapsulates both leadership and selfishness. Certain individuals indeed embody similar economic figures, leaving indelible marks on the financial landscape. That extend beyond Africa and it then adapted to be used in Africa and finds resonance especially within authoritarian regimes.

The dream of prosperity and the allure of luxurious lifestyle resonate across Africa. Indeed, the millennial culture continues to evolve and shape the zeitgeist. Economically inequality persists, it is likely to cast its shadow across centuries.

There was a time when I found myself saying that America should take over Zimbabwe politically. I can't clearly remember what led me to say such a thing, but for some reason, I did. Perhaps it was driven by economic frustrations, or perhaps by something deeper. Could it be that I resented my own people because of the past? I recall several occasions when others tried to reconnect me with my fellow Zimbabweans, but those attempts never worked. I often felt more at ease in isolation than in their company, for various reasons, especially when judgment toward others seemed to take precedence over understanding. To this day, I struggle to trust my own people. And now I see why our economy stalls, we fail to stand as one, choosing instead to drag each other down.

Economically, the West and China hold sway over Zimbabwe, drawn by its lithium, gold, diamonds, and other minerals. Yet our stagnation is not only about resources, it is also about uncertainty. No one knows how long the nation will remain misled as it is now. And so we scatter, living across the world instead of in the land of our birth. I have been asked countless times why I don't return home. That question cuts deep, and I, too, wonder why I cannot go back. For years, I have wrestled with it: why can't I find my place among my own people? Why does harmony with my fellow Zimbabweans feel so far out of reach?

This disconnection is not mine alone. There's a strange irony about Zimbabwe, even its wealthiest sons and daughters rarely live there, despite owning businesses on its soil. Take Strive Masiyiwa, for example: his family resides in the UK and travels the world, while the resources that built their fortune remain rooted in Zimbabwe. Instead of reinvesting to uplift the nation, the wealth often flows outward, leaving the soil rich but the people impoverished.

Meanwhile, ordinary Zimbabweans continue to flee in search of better opportunities, their departure resembling a quiet prison break, an escape from a system that no longer serves them. And in the stillness of solitude, a different kind of sorrow takes hold. It is not the loneliness itself, but the ache of remembering how a country once full of promise has declined in less than five decades. The fear we feel when asked where we're from is no longer about being forced to return. It is the fear of judgment. The quiet shame of being born after Rhodesia, in a time when the legacy feels more like a burden than a birthright.

Autocratic governments.

If you live in an autocratic African country, you must understand that your human rights are often limited. In places where criticizing the government or opposing its projects is forbidden, life becomes a constant negotiation with fear, fear of brutal death, fear of endangering your family, fear that silences even the boldest voices. While writing *Life Time Runs*, I often thought: if I were in Zimbabwe, even with the resources to write, I could never have told my story in peace.

Zimbabwe's struggle to defend individual rights and expose corruption runs deep. This is a country where citizens have been beaten to force them to vote, a reality as tragic as it is telling. And so the question lingers: how many of us still wonder, and how many times have we asked ourselves, what truly drives Zimbabweans to abandon the land of their birth? Beyond the headlines and the surface narratives lie unspoken truths, truths that weigh heavily on those who leave, and perhaps even more on those who stay.

At times, it feels as though a curse hangs over Zimbabwe's government a nation once cherished now turned against its own, bleeding its people dry instead of lifting them up. Perhaps someone must speak for this curse to be lifted. At times I wonder did our ancestors do something that cast a shadow over our future, so that we, too, must struggle? Not to relive their pain exactly, but to feel it in different forms.

Africans prays their masters in Europe.

They kneel in cathedrals built from their silence,

chanting praise in borrowed tongues.

The crown that once crushed their fathers' names

now gleams with reverence, not rage.

Is it gratitude, or forgetting,

this worship of the hand that once held chains?

History hums beneath the hymns,

but who dares rewrite the refrain?

The war in Gaza has indeed captured global attention, evoking emotions, debates, and urgent calls for peace. Let us read this passage carefully. Before you rush to blurt out your opinions about human rights, let us pause and reflect. Having a very good understanding of history is essential for several reasons.

Understanding the origins of the war, especially when it has religious dimensions, its crucial. When discussing matters related to religion, especially drawing from both the Bible and the Quran, it's essential to approach the topic with respect, sensitivity, and an open mind.

I don't hold opinions or personal beliefs, especially when religion and military involved. Many who are alive today, do not have great backstop for both Christianity and Islamic religions. You need to allow your brain to intermix life of BC, AD, and modern life. When you possess a fundamental understanding, you can then articulate your viewpoints. I am not going to continue diving deep in geopolitics etc.

Recently, there has been a war in Sudan, prompting foreign countries to swiftly withdraw from the country due to the ongoing civil conflict. Yes, everyone was worried. Just the moment when Russia invaded Ukraine, so everyone was more than worried. A few years ago, Cyclone Idai triggered a massive disaster in Southern Africa, affecting hundreds of thousands, if not millions, of people in Mozambique, Zimbabwe, and Malawi. The region experienced widespread flooding and devastation, with Mozambique's President Filipe Nyusi describing it as 'a humanitarian disaster of great proportion.

It is noteworthy that when significant events occur in Europe, some African leaders promptly take action to support their European counterparts. Julius Malema once said, "*Even if I didn't do anything wrong, for no reason I am scared of a white person.*" The impact of colonization has indeed instilled a complex mix of respect, fear, and dependency within African nations. Sometimes, as you walk through the city, a quiet discomfort rises, an urge to avoid crossing paths with a white person. Not out of hatred, but from a deep knowing: the system, time and again, tilts in their favor. You feel it in the silence, in the glances, in the way justice wears a selective lens.

I'm the one who brought the words and the animation together. Though they were created by different individuals, what matters most is how this piece of art speaks my truth. I've fallen in love with it, not just for its beauty, but for its honesty.

Perhaps now, you're beginning to feel that quiet shift, the easing of guilt that sometimes weighs heavy when you pass the homeless without offering help. It's not always because you don't care. Sometimes, you simply don't have anything to give. Or not enough. And that's okay. Maybe no one ever told you that. Such feeling, this tension between compassion and limitation, it can be overwhelming. But acknowledging it is the first step toward grace.

When one walks around in any city in the world, homelessness is a prevalent issue that can be observed. Having experienced homelessness for approximately four years, I am intimately familiar with the challenges and emotions associated with being without a home. I am a product of homelessness. I endured life on the streets, begging for survival, and this experience forced me to mature at a young age. In various programs, I often found myself taking on a leadership role due to my effective approaches in engaging with participants and managing the programs.

The problem with my leadership style is that some programs have taken advantage of it countless times. I refuse to let that continue, especially now that I am educated enough to recognize and avoid such exploitation. With even the most basic skills, I can also see clearly how governments around the world treat poor countries.

Believe it or not, most of them didn't like me being the leader. Maybe it was because I often made big mistakes, mistakes that weren't just embarrassing for me, but for the entire crew I was working with. I came to realize how quickly people forget that I'm human too. Just like them. I think differently, I respond differently, shaped by my own life experiences. But that difference, instead of being understood, was often judged. Many forget that I once had no home. Every choice I make is shaped by that memory, by the quiet vow never to return to that life. I will not live it again and it cuts deep to watch another soul walk that same hard road.

Now, as I write these books, the questions keep coming. Not out of curiosity, but out of criticism. Some even try to humiliate me in public, claiming I don't know what I'm talking about. But I do. I know exactly what I'm talking about, because I've lived it. And that's something no one can take away.

I can see why they say that, the words in these books make some feel guilty. Some still carry deep hatred, others are simply unwilling to learn from their mistakes. For some, it is the pain they cannot bear to revisit, and for others, it is jealousy born of ignorance.

Culture and Marketing Adaption.

I was born into poverty, but I hold the belief that I will not die impoverished. My spirit is resilient, rarely succumbing to defeat. The sole barrier to my success is the fear and lethargy that prevent me from harnessing my fighting spirit effectively.

When my spirit faces defeat, it adapts. I embrace new concepts to get back on my feet. It's a strategy akin to foreign markets adjusting their tactics when African laws don't align with their interests. Africans have largely adopted American culture, surpassing that of their former colonizers. While one might assume this adoption is an effort to claim as their own but it's a strategy to establish a market base.

Regrettably, the majority remain consumers, failing to seize opportunities for growth and development. If you examine the Hip-Hop music industry, it's predominantly influenced by Western culture. Despite the innovations brought about by the technological landscape, I question the necessity of continuing to contribute to Western markets, especially when considering geopolitics. The more we invest in Western markets, the more we risk and undervalue our own local economies. Do you remember the question from Imbrahim Traore "*why should we continue to live under the currency that does not bring value or our market?*"

BRICS and dictatorship.

At the tender age of seven, I fled my home to brave the streets of Zimbabwe and South Africa. The journey was arduous, yet escaping an abusive family ultimately proved to be rewarding. Now, I can share my story with you and express my heartfelt gratitude to the South Africans who invested their resources and effort into nurturing me.

Now, you might be wondering, 'Why are you telling me your story instead of elaborating on the transition from bricks to dictatorship?' Africa is also known for its dictatorships, authoritarian regimes, autocratic governments, and the plutocratic lifestyles of certain individuals. Let's shift our focus more toward the aspect of dictatorship. The country I hail from is widely recognized for having one of the most enduring dictators in the world, despite the presence of authoritarian regimes run by royal families, particularly with the rest of the African region. Whenever I stroll through Cape Town in South Africa, I encounter criticism regarding my country's leadership strategy. Some label me 'Mugabe,' while others express fear that I might succeed him and perpetuate his legacy. Some have speculated that his survival was due to our shared identity. For some white individuals, every Zimbabwean they meet becomes a reflection of him. To some, the name Zimbabwe carries the weight of his racism and his deep resentment toward colonial systems and the history they represent.

You might have dogs to guard your home, but if they are not fed, they cannot perform their duties effectively. You've trained these dogs exceptionally well to fend off intruders, yet their loyalty remains unrewarded. Zimbabwe is known to treat its citizens in such a manner. The country boasts one of the best education systems in the world, yet its people suffer more than those in many other places. Government authorities often undergo rigorous training to qualify for employment.

These people will not continue to romanticize poverty. Eventually, they will resist, adapt, and seek new ways of living. To understand this shift, look to the soldiers once tasked with guarding the mineral wealth of eastern DRC. Over time, many turned to illegal mining themselves, driven not by greed alone, but by survival, disillusionment, and systemic neglect. Their actions now fuel a conflict entangled with the M23 rebel group, reportedly backed by Rwanda, whose pursuit of control over land and resources has escalated into a brutal civil war.

Somewhere and somehow Asia owns Africa

Africa hosts a greater number of Chinatowns than any other region in the world. Most of the less quality products and services usually end up in Africa while some of the plutocratic individuals enjoy better quality life in either Europe or North America. We welcome any products or services that we believe are worth the investment. Remember, some of us may not have had access to formal education, so we may not fully understand the actual effects of these products.

We often undervalue our own labor compared to that from abroad. Products made overseas are frequently perceived as having greater worth, and we place unwavering trust in them, welcoming their influence on our economy. While there is a common perception that China does not always produce the highest quality goods and services, it's important to consider the broader context and the reasons we continue to rely on them. As I mentioned, the pricing of both products and labor remains one of the most significant factors

SA is in big trouble:

South Africa cannot simply break away from the agreement it made with China, it is effectively permanent. The country owes a substantial debt, which continues to place strain on its economy. On occasions when repayment is not possible, China has reportedly sought land in exchange, and control of that land means control of vital freshwater resources essential to citizens. In my assessment, any breach of this contract could provoke military intervention, an area where South Africa's capabilities cannot be compared to, nor compete with, those of its much stronger counterpart, China.

I previously spoke about South Africa, but I neglected to mention the gold mining operations in Zimbabwe, particularly in Gweru, Kadoma, Chakari, and Golden Valley. These activities are part of a wider pattern of developments across Africa, reflected in large-scale infrastructure projects such as Kenya's modern railway system, which stands as a visible measure of progress.

On 21 May 2023, Russia donated rescue helicopters to Zimbabwe, an act I would, as a leader, conditionally support, making it clear that such support would depend on specific requirements being met. In my view, Russia's recent actions appear more supportive toward certain African nations than China's current approach.

I regard many Chinese actors as displaying some of the most overt forms of racism, doing so openly while simultaneously pursuing profit across our continent. This pattern will only change when African leaders collectively decide that enough is enough.

They have a hold over us, and we cannot simply escape that reality. I remember back in school, when we had a motivational speaker. At the end of her talk, I asked about the Chinese market flooding Africa with low-quality goods. She emphasized that China plays a crucial role in the global economy. In fact, it could be that 95% of the products I use originate from the Chinese market. The truth is, there's no realistic way we can avoid them, they make access to goods far easier for Africa than many other international markets do. It is disheartening that we remain bound to a market so damaging to our economy.

Recently, seven of them were each sentenced to 20 years in prison in South Africa for human trafficking. What's striking is the silence from the Chinese government about the events of 10 September 2025. It also makes me wonder, how many crimes do citizens of economically advanced countries commit in Africa each year, and how many of them go unpunished?

Purchase your own identity:

This subheading speaks to the reality of buying your own ID. Yes, that's right, corruption never shies away from such things, but it is the world we live in. Next time, I won't be shocked if survival itself comes with a price tag. Much like people across the world who contribute through taxes.

I grew up in Cape Town, a minor who came from the streets and eventually made it through high school. For most of my childhood, I never worried much about legal documentation. That changed the day one of the volunteers at The Homestead CYCC mentioned she wanted to take a child to America and that child turned out to be me. From that moment, everything shifted.

Life became more challenging, yet it began to make sense in a new way. My fear of being deported and forced to return to the same life I had once escaped grew to an entirely different level. Today, I have grown within a system that has shaped me in ways I could not have understood back then. I now see with greater clarity than in those earlier years. This is why I often wonder whether I would have ever received even a basic education if I hadn't escaped my abusive family in Zimbabwe.

I had to travel to Zimbabwe without my family's knowledge to obtain my documentation, as I knew that financial matters were their top priority. They would obviously expect some money from me. Before left I South Africa I paid someone in Zimbabwean Home Affairs to draft a birth certificate for me. As I got off the bus Harare, he handed over my birth certificate and we immediately headed to legal resources offices to get my ID registered and to which I was told that I am not a Zimbabwean I should go back to South Africa since I was failing to speak Shona. Of course you can imagine that became argument to which almost all the public members were listening while they were waiting for their documents.

The officials teared my birth certificate and told me to go and get my father to which is why I ended paying more to my allies again. I wouldn't disturb my father from where he was laid to rest. It took me one weekend and the following Monday I was told to go and get my ID that ended up coming out within 2 hours the day I registered.

To get the full story, read "*Life Time Runs*" on Amazon. Imagine being in my place, where in the very country of your birth, you have to pay for something as fundamental as your identity. A service meant to be free was stolen from me. On the surface, it appeared to be a Win/Win, yet inside, psychologically and emotionally, it was a Lose/Win. They walked away victorious, and I was left with the loss. That truth, I cannot deny. Life was never meant to unfold that way.

Witchcraft.

I love my mother but it became difficult to love her when she wished me to die through accident with the bus. Even if she was joking trying to change my mind to keep me so that I can make money just for her unfortunately that only grew my hatred within. I did not only disliked her, but I also feared her, I saw a killer in her. I knew that she was capable to do that, so she can benefit.

To this day, I do not want to see her or be anywhere near her. I am deeply afraid of anything connected to black magic. I believe that genuine Sangomas are now difficult to find, as most of those available are fake and only interested in money. To me, a traditional healer is meant to be a traditional healer, a doctor who helps the sick, not the kind of harmful figures we see today. My mother believes in Sangomas, but I do not share that belief.

Witchcraft exists in every corner of the world, that much we cannot deny. It has been present for thousands upon thousands of years. Visit even the oldest civilizations, and you will find someone who will speak of witchcraft within their community. Without evil, good would have no meaning. Our task is to keep pushing back the darkness, so that when the light comes, it can shine in all its greatness.

I will not claim that my mother is among those who practice it, I don't know. Even if she did, I would not judge her. Yet, her actions speak in ways that seem rooted in witchcraft.

During my years in the Western Cape, South Africa, I found myself slowly piecing together the threads of a belief system that runs deeper than most outsiders will ever understand. Around lunch tables and late-night conversations, friends and colleagues would speak in hushed, knowing tones, about jealousy that could curdle goodwill, about the invisible currents believed to stir unrest in workplaces and communities.

Some would take long journeys to the Eastern Cape, seeking the counsel of elders or healers, performing rituals they trusted to shield them from unseen harm. These were not empty gestures to them, but inherited safeguards, rituals passed down through generations, meant to keep misfortune at bay.

I heard of accidents and uncanny events that, to their eyes, could not be dismissed as chance. Whether or not these forces could ever be proven was beside the point; the belief itself was woven into the way they interpreted the world. And living among them taught me that in matters of spirit, real or perceived perception can hold as much power as reality.

I believe Africa carries some of the deepest entanglements with this practice. The way we turn on one another is on an entirely different level. Instead of learning from each other and working together to grow Africa's wealth, too often we choose to tear one another down. And when we destroy each other, we are, in truth, destroying ourselves.

Chapter 3:

Internet of things VS your African village life.

Let us proceed to build a dwelling in a rural village. For someone born in an urban environment, the transition to rural life and the task of constructing a home there, can present real challenges.

Many black children raised in cities are often reluctant to embrace the rural lifestyle. Among other Africans, refusing certain foods can sometimes be seen as a sign of pretentiousness. Naturally, no one enjoys dealing with spoiled attitudes. A person's speech or manner of walking need not reveal whether they come from a rural background.

This reluctance often stems from prior exposure to a highly technological environment, where access to resources is faster and more efficient than in rural areas. Having grown up with abundant opportunities, reliable public services, and plentiful resources, life in a technologically advanced setting stands in sharp contrast to rural living. Artificiality permeates many aspects of urban life; food laden with chemicals may be considered tastier in certain contexts, yet the experience of eating fresh, wholesome produce in a rural area can be profoundly different and inherently healthier.

Financial Vampires simplified gold diggers.

They smile to charm, but the hunger's deep,

Not for your heart, it's your gold they keep.

They drain, they take, they never give,

Feeding on others is how they live.

While I have spoken about being scammed, I have yet to share my experiences of being vulnerable to financial vampires. Regardless of our intentions, financial exchanges are inevitable. Whether with family, friends, or strangers, money will inevitably leave our pockets or accounts. Family, without question, remains our foremost responsibility, an obligation we cannot avoid. Be it supporting one relative or several, there is always someone within the family circle we are more inclined to help.

The challenge is this: how can you support someone else financially when you can barely take care of yourself? Those you left behind in the rural areas may see you as the breadwinner, yet in truth, you're not, you're also struggling to make ends meet.

Fake rape cases.

In 2017 at around 19:00 in the evening, I was sitting and watching TV when the news reported that a young man had been arrested for rape. Shortly after, another story followed about a woman who had withdrawn a rape case, confessing that she had lied in order to extort money from the defendant.

This was the most striking part. I had encountered similar cases where young women fabricated rape claims but later faltered when evidence was lacking. My curiosity about her grew, and I spent months posing as her partner, all the while seeking to draw out the truth. We spent time together as though we were dating, but in reality, we were just friends. Once she realized I wasn't wealthy, she quietly pulled away.

She would sometimes undress in front of me while ironing her clothes, yet our interactions remained limited to conversations about relationships and the perception of South Africa as a nation of cheaters. Unaware of her own vulnerability, she eventually revealed the truth about a fabricated rape case. It emerged that she was seeking financial support because she had two children. In pursuit of this, she attempted to blackmail one of her supervisors, a man she had been involved with at work. It became clear her feelings in that relationship had been nothing more than a façade.

Upon uncovering more cases, I realised this was, in essence, a form of discreet, high-end prostitution. Her apparent business model relied on blackmail, driven by the belief that she could accumulate significant wealth through such tactics. Over time, we developed a close friendship, one that endures to this day. I remain thankful that my involvement was guided purely by educational intent.

She ultimately heeded my advice after I explained the potential consequences of her actions. It is indeed disheartening to witness young women resorting to such actions for financial gain. South Africa is indeed a beautiful country with immense potential for economic growth. Its diverse landscapes, rich cultural heritage, and abundant natural resources make it a unique and captivating nation. However, realizing this potential requires effective leadership and strategic vision.

The unfortunate aspect of the story is that the same individual with whom she was in a serious relationship ended up cheating. It was indeed heartbreaking to witness her in tears. This became evident that she was depriving herself of valuable learning opportunities, hindering her growth both professionally and personally. We became personal advisors to each other, even though I am not adept at pursuing romantic relationships, and my introverted nature didn't attract any interest from women.

Now this is very bad for the economy. As Africans, we have maintained our connection to nature as an inherent part of being human rather than seeking ways to grow. We often gravitate towards the path of least resistance. When someone creates a system that offers instant gratification, we tend to flock in that direction. Tell me that there is a give away of iPhones, you will see on how I will bombard you with questions, needing the details to obtain it.

In fact, I was scammed either 2021/ 2022. I encountered a scammer who exploited my trust by using iPhone 12 to extort money through iTunes. When reflecting on the matter during discussions with friends, I often find myself expressing a preference for hackers over scammers. Hackers, although often associated with unauthorized access, typically refrain from financial theft. In contrast, scammers employ persuasive tactics to convince individuals to part with their hard-earned money, even if its a small some to keep the account active.

Now, what if I had children by then, that means they were going to die with hunger because I needed expensive iPhone. Due to my pursuit of instant gratification, I incurred certain costs.

We undermine our own progress when we rely on others to do everything for us. While residing at the children's home in Khayelitsha, we did not handle our own cooking and washing. Most things were done for us. As we immersed ourselves in that particular culture, our expectations for others to handle tasks on our behalf grew. Adapting to life in Du Noon presented challenges, especially when I had to rent a place. I found myself washing clothes in a large basin and cooking on a gas stove in my small room.

The most effective way to address such challenges is through education. Practical education in taking personal responsibilities is essential. Born free' generation enjoys more opportunities, it is crucial to engage in an informed discussions about the past.

Technology effects

Technology has indeed opened up new avenues, but understanding historical injustices remains vital for fostering empathy, promoting reconciliation, and ensuring a more equitable future. I do recognize the historical context and the disadvantages caused by technological advancements. I am actively working to combat the inertia and complacency that have taken root within our community."

Financial management and investing in tech education are critical components for personal and societal progress. Learning from past technological errors is crucial as we now witness the emergence of Artificial Super Intelligence (ASI).

Now, let's explore the impact of the internet. The spread of false and misleading news occurs much more rapidly than the dissemination of accurate information.

Criminals often target crowded networks due to the allure of drama, a phenomenon observed not only among Africans but all around the world, especially Western society. Taking action that capture public attention and trust me it can have a global impact within minutes, thanks to the internet and advanced technology. While the internet enhances our lives in many ways, it is essential to acknowledge that there is a significant downside to its widespread use. We often use this negatively than we realize, even when our intentions are good. The original purpose of the internet was primarily for communication and business, rather than the diverse range of uses we see today.

Rich keeps getting richer.

The situation appears to be deteriorating. Could you please ask the following questions? The potential impact of technology replacing humans in the workforce is a significant concern? Will the wealth continue to influence our decision even further? Will humans turned into bots or are we going to lose our individuality? Besides helping humans remember and acquire information at a faster pace, what are other aims of technology like Neurolink? If someone were to hack into the chip implanted in your head, several potential consequences could arise? How does that have influence on Africa's Lost Wealth?

When examining the questions raised, it's evident that we often grapple with a sense of limited control. Self-control is also just another state of mind just like the word freedom. Will not being in control of our own lives in today's world governance. Having advanced tech will further have negative influence on the Africa's Lost Wealth? Unless we establish a significant presence in the tech industry ourselves, we will remain dependent on external forces. Otherwise the rich will keep on getting richer. These situation will only continue to escalate if we allow it to persist.

Think of the colonization. The astute colonizers strategically controlled access to a vital resource, allowing them to amass significant wealth through both religions and advanced tech. Its only matter of two to three years they will then start depopulating Africa to recolonize it again through testing Neurolink on Africans since this continent has been their testing laboratory. In fact, it is already happening.

If you visit Kenya, you won't find the Mid-Atlantic Ridge, that lies beneath the Atlantic Ocean, but you will witness another extraordinary geological phenomenon: the **East African Rift System**. This vast rift is the result of a divergent boundary between sections of the African Plate, specifically the Nubian and Somali plates, which are slowly pulling apart. Over millions of years, this process could split Africa into two separate landmasses, creating a new ocean between them.

This raises a pressing question: will such a division further erode the wealth and unity of the continent, or will it reshape Africa's geopolitical landscape in ways we cannot yet predict? What's certain is that the fracture is progressing from the eastern side of the continent, with profound implications for borders, resources, and alliances.

The **Garden Route** in South Africa, a fertile, biodiverse stretch along the southern coast, is a reminder of the natural treasures that bind us. Preserving such regions is vital for our unity. In the event of a continental split, new laws and power structures will almost certainly emerge, shaped by a mix of military strength, technological capability, and geopolitical influence. It's no secret that Western powers have long had strategic, and some have already interests in Africa drafted policies anticipating shifts in the region's political and economic order.

Africa's true scale is often underestimated. In reality, it is so vast that the United States, Europe, China, India, and Japan could all fit within its borders with room to spare. This size, coupled with its resources, explains why external powers have historically sought and continue to seek, a foothold here.

From a scientific perspective, Africa is the cradle of humankind. Genetic and archaeological evidence points to southern Africa, particularly regions of modern-day Botswana, Namibia, and Zimbabwe, as the likely homeland of our earliest ancestors. For some, this reinforces the belief that Africa could be the biblical Garden of Eden. While theologians and archaeologists continue to piece together evidence, the symbolism remains powerful: Africa as both the origin and the inheritance of humanity.

Social media robs you, your time.

"*Time is like flowing a river, you can't touch the same water twice. Use it wisely since it is the most valuable currency than the perceived value known as money*"

Every day I make sure that I spend time on my social media accounts. While writing this book, I had times where I would spend time on my social media advertising this same book even before I wrote it. Most of my business mindset came from the influence of spending time on social media because I wanted to make money and buy the latest version of the flashy materials that I either saw on Instagram or any other social media apps.

Back in high school, we occasionally had a motivational speakers. They would discuss about the consequences of spending time on social media during their speeches, but interestingly, they would also mention their own social media accounts for a follow-up. As we return to our class after listening to this individual, I would summarize his message as an encouragement to remain mindful of our social media usage.

Some of my classmates would occasionally inquire about the meaning of the speech, as they considered it to be the best. I would often remind them that the speaker encouraged us to follow him on social media, implying that we should continue using our social media platforms. From my perspective, it was a valuable experience that not only provided inspiration but also sparked some business ideas. You also need to remember that if you feed a beast, that same beast is going to kill you. As a child who fled Zimbabwe, my story might lead you to assume that I grew up without certain privileges. It's true that tough times are a part of life, but fortunately, they don't last throughout the entire year. That gave me a purpose to live by chasing my dreams including becoming a businessman. Those days I had spent on social media, I had pushed back more than needed time to spend chasing my dreams. I also became in secured, got sick and I developed poor mindset of having so much of wishful thinking. These were the results of witnessing someone else's blessings.

What's striking about my story is the blame I keep placing on my past, as though it alone explains my failures today. This mindset is not mine alone, it's a silent epidemic in many corners of the world. It's a kind of mental sickness we rarely notice until it has already crept into our thinking.

I understand all too well how difficult it is to release the grip of a negative past and keep looking forward toward something that brings true growth. But we must choose to stand for that growth.

As I wrote in Chapter One, the African story, though scarred by endurance through centuries of hardship, has also given rise to great opportunities for its people. These opportunities emerge despite the systems that still restrain our collective economic.

So let's focus on what we can change. Where change is not within our reach, that is where we must come together in African unity. Only then can we build an economy strong enough to stand on its own, rather than wasting our energy lamenting its failures.

Fantasy to reality.

I once saw a question on YouTube: *Why do billionaires work so hard to look like everyone else?*

At first, it felt like an observation about wealth and disguise. But the longer I thought about it, the more it seemed like a mirror, not just for the elite, but for all of us. In different ways, we too try to blur our true identities, bending ourselves into shapes we think the world will accept.

Before we go further, recall the heading of this paragraph and remember what I wrote before it. There, I spoke about the time I spend on social media each day. What began as an intentional immersion in technology gradually became something else, a slow drift into addiction. My attention shifted from building tools of progress to curating images of myself for an audience I could not see, but somehow feared to lose.

The wealth portrayed on social media is a blend of both genuine affluence and artfully crafted illusions. Everywhere in the world, there are individuals who conceal their true personalities. In the realm of social media and business the phenomenon remains consistent. In the business landscape, one encounters a spectrum of entities: authentic businesses run by genuine individuals, alongside con artists and those who aspire to transition from being fake personas to legitimate business owners. The line between authenticity and deception has become increasingly blurred, making it challenging to discern what is genuine and what is counterfeit.

Despite the drawbacks of social media, I have gleaned valuable lessons about both people and the business world. My exposure to social media has opened my eyes to numerous opportunities and imparted essential life lessons.

That question about billionaires trying to look like everyone else is not just about the wealthy, it is a quiet allegory for Africa itself. For too long, we have been told to mirror the image of others, to dress our economies and cultures in garments stitched elsewhere. In chasing validation, we risk losing the texture of our own identity.

And yet, the strength we seek will not come from imitation, but from authenticity. Just as I had to recognize when my energy was being drained by the wrong pursuits, Africa must turn its gaze inward, valuing its own patterns, its own rhythms, its own genius. Only then can unity become more than a dream, and economic strength more than a wish.

Get rich quick scheme.

A truly wealthy individual has mastered the game. A truly affluent individual understands the deeper intricacies of building and sustaining wealth. *Wealth* and *richness* are often used as if they mean the same thing, yet their meanings are distinct and most of us fail to see the difference. Remember this: a rich person still worries about money and the struggles of day-to-day life. In that sense, they are not so different from the poor.

A truly wealthy person knows how to leverage their resources to create opportunities and build lasting prosperity. *Richness* often refers to having substantial financial means, but it doesn't necessarily ensure financial security. A rich person may still encounter financial challenges even periods of being broke. For the truly wealthy, however, "being broke" is either a relic of the past or a moment they can turn into an opportunity for greater accumulation.

I became hooked on YouTube videos about billionaires and millionaires. Even as I was writing this book, I had countless clips downloaded on my phone to watch during my off days. In my quest for a quick path to wealth, I turned to YouTube for answers. But I soon discovered that true wealth demands more than shortcuts, it requires diligent research, strategic investment, and the building of tangible assets. At times, I even tried my luck with gambling, hoping for an instant breakthrough. Yet it didn't take long to realize that relying on chance alone is no foundation for lasting success.

True prosperity demands a deliberate approach, grounded in informed decisions, prudent investments, and a steadfast commitment to building lasting wealth. I immersed myself in the lives of self-made young millionaires and billionaires, searching for clues to their success. By studying their strategies, work ethics, and investment choices, I hoped to uncover lessons that could accelerate my own journey. Yet this pursuit carried its own danger. Rather than acting on tangible opportunities, I became tethered to my device, chasing shortcuts and quick fixes. The seductive promise of instant success kept me staring at screens, while the real world and the work it required passed me by.

The opening statement of this subheading declares that a wealthy individual has *mastered the game*. With significant wealth often comes influence and, in many cases, control over others. This power can create a dynamic where some people become subservient, setting aside their own dreams and aspirations to serve the interests of the wealthy lifestyle.

How do the wealthy maintain such influence? One avenue is through the creation of employment opportunities, which can benefit both individuals and governments. By generating jobs, organizations fuel economic growth and stability, while governments benefit from increased tax revenue and social cohesion. It is a symbiotic exchange, where financial success and societal well-being appear aligned.

However, the platforms they create are not limited to traditional employment. Some also establish ventures such as gambling enterprises industries that, in Africa, find a particularly eager audience. A closer look reveals that Africans make up a significant share of users in this space. Savvy foreign tycoons see the immense potential of these markets and are increasingly embedding themselves in business ventures across the continent.

The inclusion of this subheading is fitting, as *Get-Rich Schemes* proliferate across the internet. Over time, the influence of the *Internet of Things (IoT)* will only grow more pronounced. Elon Musk has predicted that people will increasingly connect to the internet to think, communicate, and travel more efficiently. Yet we know that such optimism will not always match reality. While technological progress continues to enhance our lives, it also brings security vulnerabilities that must be addressed. In every human-built system, side effects and unintended consequences are inevitable and they demand our attention.

New startups VS old business strategies.

Starting a business in Africa demands resilience, adaptability, and a deep understanding of local realities. Entrepreneurs who embrace these challenges have the potential to drive economic growth and leave a meaningful imprint on their communities.

Yet the modern startup landscape carries its own pitfalls. Too often, founders lose sight of their original vision as they navigate the journey. After securing funding, some make questionable choices about where to invest, and even when they achieve a return, they may misuse profits resulting in outcomes far short of their potential.

For the past six years, I have worked at V&A Waterfront's Battery Park, a vantage point that has given me a sharp eye for the ebb and flow of people in the area. I can tell who is new, who is established, who has real financial affluence, and who merely appears to.

Observation goes beyond the surface. I've seen how people manage their money, some squander it, and others invest with foresight. My own financial decisions have not always been wise; some have placed me in deeply uncomfortable situations, shaping the life I lead today.

From my post at Battery Park, I have often watched people younger than me drive the latest model vehicles I could never purchase without a lifetime of saving. It is tempting, in such moments, to surrender to the belief that such luxuries are forever beyond reach. But the voice of my mentor returns to me: material possessions will always exist. The task is not to chase every glittering thing, but to work relentlessly toward what you truly need, the kind of goals that outlast mere desire.

This period has given me the chance to rise toward a brilliant new idea, to invest in sectors that have a tangible, lasting impact on society: land, transportation, agriculture, healthcare, education, housing, and technological innovation. It is disheartening to see resources squandered on fleeting indulgences when they could be channeled into endeavors that outlast the moment and serve a greater purpose.

It's not that I see myself as superior to those who hold wealth; the difference lies in my roots. Growing up with limited means, taught me the discipline of allocating funds wisely, choosing essentials over perishable luxuries.

I know the weight of financial insecurity, how it can feel crushing, especially when measuring our lives against those who seem more stable. But every path is different, and success is rarely a straight line. The rises and falls are not interruptions to the journey, they are the very substance of it.

Financial freedom is not merely about accumulating wealth, it is also about peace of mind, a sense of security, and the ability to make choices that align with your values. Every step forward brings you closer to those goals.

While Africans understand this truth, many carry the deep imprints of a history in which their forebears endured the brutality of colonialism. Such inherited memories, though powerful, can sometimes cloud the path to self-development. We must be careful not to allow the experiences of older generations to become an unchallenged blueprint for our own lives.

History is a valuable teacher, yet not every lesson from it serves our present growth. Applying these boundaries is difficult, especially when patience is demanded beyond what is reasonable. But discernment is essential for only by balancing respect for the past with clarity about the future can true financial and personal freedom take root

Imman Ghadzi: said that "*When new opportunities arise, there are often several more just around the corner. This abundance can make it difficult to choose one to commit to. The tricky part is this: if you don't know how to choose, it may be because you haven't yet discovered what you truly want to do in life.*

On the other hand, starting something only to abandon it the moment it becomes too hard and then jumping to the next opportunity, can also make the journey more difficult. Growth requires persistence. Hold on to one pursuit until you've mastered it or achieved your intended outcome, and only then consider moving on to the next."

Churches are business.

Beneath the steeple's solemn grace,

coins still change from hand to hand.

Prayers rise, yet in their place,

ledgers grow across the land.

Throughout history, churches have not only been centers of worship but have also operated with many of the characteristics of businesses. Even in ancient times, religious institutions played influential roles in society, often intertwined with economic life.

During African colonialism, religion assumed a complex role from missionary activities that sought to reshape cultural identity to the imposition of colonial currencies that altered local economies. Yet, amid these historical shifts, selflessness, as rooted in biblical teachings, has remained a defining virtue for Christians, a call to love, serve, and uphold humanity.

Today, faith practices have branched into diverse expressions, shaped by the evolving values, challenges, and perspectives of our generation. The essence, however, still rests in the tension between spiritual conviction and the realities of an ever-changing world.

Some individuals in positions of religious leadership have used churches as vehicles for personal wealth accumulation, amassing riches while maintaining the outward appearance of spiritual authority. Some individuals also engage in acts of racism, whether through discriminatory language, exclusionary behavior, or biased decision-making. When people place their trust in such leaders for guidance and moral direction, financial exploitation can cause deep harm. Congregants often give through tithes, donations, or support for church projects, in the sincere belief that their contributions will nurture their spiritual growth and serve a higher purpose.

This is yet another way in which African wealth being stolen in plain sight. Funds that could have fueled community development, reduced poverty, advanced education, and strengthened healthcare are instead diverted elsewhere. Resources that might have built schools, hospitals, and vital infrastructure remain locked away through mismanagement, placed in the wrong hands, and thus denied to the people they were meant to serve.

If we continue along our current trajectory, when can Africa's economy realistically achieve self-sufficiency? As we navigate the competing influences of both Western and Eastern economies, the long-term implications remain uncertain and complex.

Why does Africa continue to anchor itself in the tangled economic webs of the West and the East? What is it that we truly want? Is there a collective prize at the end of this path, or is this alignment merely a classroom where our leaders learn new ways to tighten their grip on our societies?

Take a moment to watch the growing number of videos exposing how some African pastors extract money from their congregations. The reality is both absurd and heartbreaking, absurd because of the theatrics and false promises, heartbreaking because the very people being exploited are often struggling to make ends meet. These churchgoers place their trust in leaders who were meant to uplift them yet instead find themselves victims of manipulation and financial harm.

Please lead us in the right way.

It is essential to cease the suppression of innovative individuals, thereby fostering a robust and flourishing economy. Despite fleeing an abusive family in Zimbabwe, I have resolved not to return due to the country's struggling economy, which shows no signs of improvement. The economic decline in Zimbabwe can be attributed to the current leadership's approach and governance style. The enduring impact of Mugabe's leadership continues to influence the current state of affairs.

Nations worldwide are striving to emulate the economic model of Finland, renowned for its stability and innovation. We all aspire to reside in a society that mirrors the prosperity and advanced infrastructure of the first world. To achieve such a collective goal, it is crucial to establish a unified agenda that aligns with our shared objectives.

So far, only a handful of African leaders have reached the resolute conclusion that enough is enough. They have chosen to reject the Western leadership systems that, under the guise of guidance, have long oppressed African societies. Among them is one of the youngest leaders, Ibrahim Traoré, whose decisive leadership has been evident from the moment he took the oath. Since then, he has driven economic progress in his country at a pace surpassing that of leaders who have held power for decades.

Today, many leaders remain hesitant to lead with the courage and conviction shown by Ibrahim Traoré. The question before us is: what motivation can we offer to awaken that same fearless spirit within them? And what systems can we put in place to guide them toward becoming truly great leaders? We do not seek to remove them. We recognize that they hold valuable insights into the economy and the path toward a unified Africa. Their knowledge is an asset we cannot afford to lose. What they lack is not vision, but confidence

Dark side of the music industry in Africa.

The beat was born in the soil,

carried by calloused hands to the mic.

But in shadowed boardrooms,

golden voices are sold for copper coins

and the drum still pounds,

though its rhythm is caged.

I have been living in South Africa for nearly 18 years now. Through both teaching and learning, I have gained valuable insights, particularly in the realm of identifying and assessing individuals. As an orphan, I grew up surrounded by a diverse array of personalities in the children's home.

That exposure to a wide range of personalities have strengthened my judgmental skills. In different areas, there existed a cultural habit of watching music videos. In Khayelitsha, we predominantly enjoyed House music, while in District Six, our musical preferences shifted towards hip-hop. Despite my occasional pretense, I found myself influenced by the desire to emulate African American culture. During that time, I uploaded some YouTube videos where I attempted to rap and follow the beat.

Allow me to share a brief anecdote drawn from my biography titled "Life Time Runs." This personal experience serves to underscore my point.

The influence of media is far more potent than we might imagine. During my years in the children's home, we would watch popular soap operas such as 'Generations: The Legacy' and 'Muvhango' etc. After watching those programs, we would often switch channels and either enjoy soccer matches or watch hip-hop music videos. Watching hip-hop music videos had a significant impact, especially during the period between 2012 and 2019, which coincided with the rise of hip-hop culture in South Africa.

Now, hear me out: I recognize that I was once a coward. As I attended school every day, I increasingly questioned the benefits of such a lifestyle of Hip-hop music. Many of my peers who immersed themselves in hip-hop culture ended up developing arrogance rather than maintaining discipline to pursue a music career. One individual has returned to the streets, while the whereabouts of the others remain unknown. While I recognize that I have no right to judge, it's disheartening to witness the untapped potential of many Africans who face limited time in their lives. Social media has significantly impacted Africa's economic landscape, both positively and negatively.

In the hip-hop music culture, the very individuals we look up to often lead lives that diverge from the aspirations we hold. Behind the glitz and glamour, there exists a reality that contrasts with the image projected by these artists.

Becoming a hip-hop musician often entails a complex journey. While the genre demands discipline and creativity, it also exposes individuals to potential pitfalls. Many aspiring artists find themselves grappling with the allure of substance abuse, alcoholism, and involvement in criminal activities, as the hip-hop culture can be both inspiring and treacherous.

In the pursuit of our goals, we often find ourselves struggling to maintain focus on our studies. I, too, am guilty of this lapse. Since I became deeply involved in helping strangers accumulate wealth by liking their posts on Facebook and Instagram, my focus on school has waned. Subsequent to my increased focus on social media in trying to amass significant wealth through get rich quick scheme my academic performance began to decline, resulting in failing grades.

We must regulate and filter the sources that shape our thinking. Not everything we learn or everyone we admire, is worth following. Sometimes, the people we call role models turn out to be poor leaders in life.

For example, I was once a passionate fan of Hip-Hop music. Over time, I noticed a troubling pattern: many of these artists either die young or end up in prison, often due to gangsterism, alcohol, or substance abuse. In countless music videos, such self-destructive habits are glamorized with artists boasting about money, objectifying women, and promoting the illusion that real success is about living recklessly and getting whatever you want.

Jack Ma and Elon Musk's views on AI its advancement.

Jack Ma, the co-founder of Alibaba Group, does not believe that AI will become significantly smarter than humans or take control over them. His perspective aligns with the idea that AI should serve as a tool to enhance human capabilities rather than replace them. He also acknowledges the visions that Elon Musk have on the development of technology and space science. "*Humans have different personalities (characteristics)*". Jack Ma believes AI can excel in specific domains; it is unlikely to surpass human intelligence in the foreseeable future since it is a human invention. He further jokes saying that I don't want to go to Mars because I just came from there.

On the contrary, I agree with Jack Ma's assertion that humans cannot and will not create another human. At our current level of technology, the closest we can come is 3D-printing materials that only resemble human skin. Beneath the surface, however, it is still just plastic.

As someone who was born and raised in the Christian faith, I share the belief that humans do not create other humans directly, as we all trace our origins back to a single God. Jack Ma contends that those who believe technology is inherently smarter are often individuals with college education, comparing them to those with street smarts.

When you commit yourself to relentless hustle and the active pursuit of your goals, you often gain lessons far richer than those acquired by most who pass through university lecture halls. At the heart of all things lies a spiritual essence. Yet, paradoxically, human beings often understand machines more deeply than they understand one another. Or perhaps it is more accurate to say that those who own technology companies, along with those living in urban environments, are the truly fortunate ones able to fully understand and benefit from technology far more than those in remote areas. Machines remain inert until awakened by human command, while the inner workings of our fellow humans often remain far more elusive.

Elon Musk, on the other hand, disagrees with Jack Ma's stance, yet in some ways works in parallel with him. Musk has stated that, despite his differing views, he believes love always wins not just any love, but a spiritual love that can only be given and received between human beings. Such love, he suggests, ultimately prevails over all else.

It is, however, surprising to hear this from Musk, given that his public persona often mirrors the stoic, hyper-pragmatic mindset of figures like Andrew Tate, who openly frames love as a vulnerability rather than a strength. Tate rejects emotional attachment, insisting that a man should remain in control, unshaken by sentiment.

This clash of philosophies reveals a deeper battlefield: on one side, the belief that human connection and spiritual love are the ultimate forces shaping our future; on the other, the conviction that emotional detachment and self-mastery are the keys to power and survival. It is a contest between those who see love as the highest form of human strength, and those who see it as a strategic liability, a debate that mirrors the broader tension between empathy and dominance in the age of technology and influence

I am also a big admirer of Elon Musk, yet I must remember that I am a Christian. He owns a company developing brain-chip technology, and as a believer, I cannot ignore the Bible's prophecies about the mark of the beast, the new world order, and the ambition of one person seeking to control the world. This reminds me that even as a fan and a tech enthusiast, I must carefully monitor not only what I learn, but also how and from whom I learn it.

NB: God is the only Creator, and because of Him, I will always succeed in whatever I set out to do. I depend on Him for everything I need and bring all my requests to Him in prayer (I am a Christian).

Two well-known atheists, Elon Musk and Jack Ma, share certain qualities that I also see in myself. I am creative, capable of writing books like this, and a passionate tech enthusiast. I even dream of living on Mars one day. While I am Black and was born in Africa, I share a unique connection with Elon Musk he, too, was born in Africa, is a tech visionary, and has an ambition to colonize Mars.

I lived on the streets for a little over three years, and it was there that I learned the discipline that still guides me today. I have clear goals of owning businesses in the future. I believe that formal education is not the only path to gaining the knowledge and wisdom needed to lead or serve society. Jack Ma also speaks about the value of "street education" over purely formal learning. He believes that today's formal education is not the only way to become truly educated and to live life to the fullest and I agree.

I don't need to present evidence, because it exists all around us, unless you live among the elite. There are people with tertiary qualifications who are still unable to find jobs that allow them to take care of themselves and their families. This reality proves that education, in its truest form, goes far beyond the classroom.

Now, here's what we can learn from their discussion: despite the ongoing debates, none of them are addressing the significant impact that AI will have on Africa. This suggests that resources will always flow to and from the point of creation, regardless of what that creation may be. I would not be surprised if, in the future, Africans find themselves governed through AI systems controlled by either Western nations or first-world countries in the East.

In Africa, we are evidently moving toward a future where the use of chips may become mandatory, similar to how the vaccine was implemented. While many remain skeptical about the Neuralink chip, their defensive nature is indeed warranted. It is preferable to wear something externally rather than having it implanted inside your body. If you ask the Christians, they can confirm that the number 666 is associated with the mark of the beast mentioned in the bible, particularly in the book of revelation. Which describes apocalyptic visions related to the evil and opposition to God in the last days.

They will affirm that the Bible teaches God does not desire anyone to seek control over his people. Of course, God will never allow someone else to take ownership of what belongs to him, just as no one would willingly hand over something precious to them. History shows, however, that the Bible was also used as a tool during the colonization countries. In parts of the Eastern Cape of some African, you can still find historical buildings from that era, structures that were never demolished and now stand as heritage sites.

There are certain things whose impact depends entirely on your intentions in using them the Bible (religion), money (trading systems and economic pathways), and technology (power and information). Each of these can be used for either good or harm. What matters most is your purpose, and how and when you choose to use them. They can be tools to colonize and control a society, or they can be harnessed for the betterment of humanity. Sadly, when they are used for good, those efforts are often outnumbered by those who use them in the opposite direction.

Fight a good fight.

Its time to build Africa. Its time to take over the businesses and being in control of what actually belongs to us. The question is how do we fight that and what is a good fight? Fight a good fight don't involve military intervention.

My entrepreneurial mindset and innate skill for identifying business opportunities drive my approach. I firmly believe that success lies in understanding market dynamics, seizing strategic moments, and leveraging my instincts to create impactful ventures. This experience is not intended to solve my personal struggles, but rather to prepare the coming African generation to lead the continent with prominent leadership skills. I aspire for a future where African communities prioritize interdependency over individual wealth accumulation, fostering a generation of leaders who will steer the continent toward collective prosperity. I am not promoting this book; rather, I aim to raise awareness about the misleading systems that have been implemented. Employ the most prominent strategic investment to claim African wealth to its owners without military intervention.

Navigating life's labyrinth; especially when that journey begins in a foreign land, can be both daunting and transformative. My own path, set in motion at the tender age of seven, is woven with a tapestry of experiences, challenges, and resilience.

Do I always practice what I preach? The honest answer is: only sometimes. Yet my books stand as living evidence that I strive to do what I say. I am an organic African, carrying an unyielding spirit that seeks the natural paths laid out by my destiny. I believe destiny can be delayed, but it can never be denied. I am of Egyptian Negro heritage and Alkebulan ancestry.

And so, I call this pursuit a *good fight,* one that rejects violence, transcends race and culture, and embraces pure inclusivity. It is a fight rooted in the shared striving for common goals, where unity becomes the truest measure of victory.

To be continued.

Africa's Lost Wealth

My past has imparted important lessons that remain deeply ingrained within me. Among those lessons is the importance of always saying facts. When I speak these truths, I lose people. My words stir anger, provoke discomfort, and sometimes expose the hidden colonial agenda. And almost without fail, I'm told, "You don't know what you're talking about."

This book provides insightful exploration of historical facts. offering a comprehensive understanding of Africa's past and its implications for the future of the continent.

What's even more remarkable is that such organic facts cannot be found anywhere else. The type of governance that allows someone to write and publish something does not always permit the discussion of certain sensitive topics. Some educational systems in Africa today do not have such open discussions due to the intention of breading employees.

Africa's lost Wealth is a remarkable book because it applies to all educational and economic levels, making it accessible and relevant to a wide audience. A lot of individuals have mentioned to me that I am skilled at teaching, indicating that it may not necessarily involve solely lecturing in front of a classroom. I wonder if this might be the most effective way to share what I know. Someone believes that possessing knowledge is crucial, however, I argue that the real importance lies in implementing and utilizing that knowledge.

Many Africans are passing away with their knowledge. Rather than dwelling on the past, we should take action and then provide feedback based on the outcomes. When someone inquired about my actions conflicting with my words, I responded that my book serves as a way for me to practice what I preach.

It is necessary to recover the riches of Africa, but we must acquire them through economic means rather than military intervention. The issue lies within our investment strategies. Altering our strategy towards the type of investment is more important than continuing to adhere to outdated Western systems.

Do not try to accumulate riches as a continent by exploiting religion, as faith cannot be purchased with money. Introducing religion into the pursuit of wealth will only further deteriorate our existing economic condition. Ceasing efforts to solve problems in foreign countries and instead concentrating on our natural resources, while also investing in a highly efficient secondary sector, could significantly boost economic growth in Africa.

Obert Makaza

About the Author

My name is Obert Makaza. I am from ZW originally but i was raised and went to school in South Africa, Western Cape, in Cape Town. I am the writer of this book. I hope you enjoy and I am prepared to hear your great feedback.

www.ingramcontent.com/pod-product-compliance
Lightning Source LLC
LaVergne TN
LVHW090418210926
848480LV00002B/610
* 9 7 9 8 2 3 2 1 9 5 8 8 5 *